Thomas Clifford Allbutt

On Visceral Neuroses

Lectures on Neuralgia of the Stomach and Allied Disorders

Thomas Clifford Allbutt

On Visceral Neuroses
Lectures on Neuralgia of the Stomach and Allied Disorders

ISBN/EAN: 9783744693295

Printed in Europe, USA, Canada, Australia, Japan

Cover: Foto ©berggeist007 / pixelio.de

More available books at **www.hansebooks.com**

ON

VISCERAL NEUROSES;

BEING THE

GULSTONIAN LECTURES

ON NEURALGIA OF THE STOMACH AND ALLIED DISORDERS.

Delivered at the Royal College of Physicians, in March, 1884,

BY

T. CLIFFORD ALLBUTT, M.A., M.D. Cantab., F.R.S.,

FELL. ROY. COLL. PHYS.,

CONSULTING PHYSICIAN, LEEDS GENERAL INFIRMARY, AND LEEDS HOSPITAL
FOR WOMEN AND CHILDREN, ETC., ETC.

PHILADELPHIA:
P. BLAKISTON, SON & CO.,
1012 WALNUT STREET.
1884.

TO

MY OLD FRIEND,

WILLIAM HOWSHIP DICKINSON, M.D.,

I Dedicate

THIS LITTLE BOOK.

MEANWOOD, *June*, 1884.

PREFACE.

GENEROUS as was the reception of my March Lectures by the College of Physicians, and afterwards by the profession at large, yet I did not readily believe they had sufficient permanent value to justify republication. From all quarters, however, the demand for their re-issue has been so great and so kind that I am led to hope the thoughts I so diffidently expressed may still be of some real service to my fellow-workers. The almost enthusiastic welcome of my remarks on gynæcological subjects, not only by the other branches of our profession, but also by those gynæcologists whose good opinions I most value, has been a peculiar relief to me. From all directions, by word and letter, I have been assured that my zeal for the high repute of that invaluable department of Medicine has been seen and valued by those gynæcologists whose withers are unwrung. But London practice is unhappily, at present, eaten up by specialism outspecialized. A smooth-tongued and audacious gentleman needs but six months' practice in the manipulation of some endoscope or other to become a dexterous specialist and a thriving tradesman. How is the public to judge between this man and those learned and capable observers who have found the infinite in the least of things, and who breathe into the study and teaching of the meanest of phenomena a spirit bred in the contemplation of the most lofty.

For, indeed, the best men of the day are too lenient in this matter. A young lady told me but lately that a London physician of this strength—one who is chiefly known as a "lung doctor," but yet one under whom I would gladly place myself or my friends for any malady—said to her that having found her lungs to be healthy he would name another physician who would care for her "general health." Now is this sort of thing, of which we hear every day, a tribute to an imbecile kind of etiquette, or is it an indulgent tolerance of something short of rectitude? In any case, if it continues to find encouragement, the reputation of our London schools, so dear and precious to us all, must assuredly wane. The robust general practitioners who do possess a wide knowledge of their profession, have hitherto tolerated these manœuvres to please their patients, but they are getting tired of the fashion, and contemptuous of its children.

The specialist in his turn is beginning to ignore the general practitioner and to annex his patients, and thus science loses, practice loses, and the good-fellowship of a great profession is dissolved.

Finally, in giving this little book to the world, may I beg my readers and critics for a moment to consider my purpose in so doing. For instance, it is no part of my purpose to discuss at length any other disorders of the viscera than the neurotic; still less to ignore the existence of the many morbid states of them which take their beginning in tissues other than the nervous. To say, therefore, that I propose to sweep all visceral disorders into the net of neurosis, or that I see in all of them this single feature to the exclusion of the rest, is to overlook not only the confines of my subject, but also much of my incidental teaching.

I have even urged that in neurotic persons it is especially important to search for any kind or degree of local offence which in them may have aroused nervous reactions so imposing and distressing as to conceal the original seat of the disturbance, and to establish a secondary malady out of all proportion to the mode of its initiation. Peripheral irritations, in a word, may have no special alliance with the nervous system, or, on the other hand, may be either the cause or the consequence of nervous perturbation. It is of these latter conditions only that I have undertaken to treat.

P.S.—After the revision of these sheets, I received the gift of his "Diseases of Women," from Dr. Macnaughton Jones. The preface to that work, written as it is by so able and learned a gynæcologist, is a pillar of support to those who hold like opinions on the mischief of modern specialism.

LECTURE I.

Gentlemen,—When the revered chief of this College bade me to the high place which I now hold, I would fain have been released from so great a call. It seemed to me that the leisure in which I should search out and think things worthy of so distinguished an assembly had long been denied to me. But I was assured that, although lectures containing the results of systematic research are of the highest importance, there is a place, nevertheless, for lectures conveying the humbler reflections of working physicians who deal with the daily experience of the sick-room and the consulting chamber. It remained for me therefore to obey, and to deal with some part of the field of practice. Herein I have chosen for my subject those painful disorders which have their seat in the nerves of the viscera, but I have entitled my lectures the " Neuroses of the Viscera" in order that I should be free to enlarge occasionally upon some perturbations of those nerves which are not attended with pain, strictly so called. At the same time, it is with such neuroses as are manifested in pain that I intend more especially to deal. It seems to me that in our textbooks these affections receive something less than the full treatment which their difficulty and their importance demand, and I find in general practice that the part played by the nervous system in visceral disorders is insufficiently seen. To the knowledge and skill of physicians so highly placed as yourselves I cannot hope to add anything, but I would rather seek, through you and with your aid, to spread abroad a more adequate interpretation of the disorders in question, saying to you, in the words of the Apostle: "Οὐκ ἔγραψα ὑμῖν, ὅτι οὐκ οἴδατε τὴν ἀλήθειαν, ἀλλ᾽ ὅτι οἴδατε αὐτήν."

It certainly happens that neuroses above the belt are more clearly understood than those below. This preference is due in part to the renewed impulse given to the study of thoracic disease by the discoveries of Laennec, partly to the more vivid functions of these parts, functions which excite our admiration and interest in greater measure than the brooding and silent life of the organs of vegetative existence. Of the neuroses, then, of the upper chambers of the body, of migraine, of asthma, of angina pectoris and the like, I shall find but little to say; my attention will be confined almost entirely to the pains and storms of the abdominal regions—to gastralgia, nephralgia, hepatalgia and their allies. Of the neuroses of the pelvic viscera much has been said. Indeed, if thoracic neuroses have been discussed at their actual value, if those of the abdomen have received less than their due meed of attention, surely those of the pelvis have received an attention, if not beyond their importance, at any rate of far too exclusive a bearing. What I say of these last will be little in description, but more in the way of interpretation.

The first strong impression which I received from the facts of the class with which we now concern ourselves was obtained from the index of my case-books. It has been always my custom to index therein the names not only of my patients but also of their maladies. I have to each volume two indexes, one of names, one of diagnoses.

As each year I have prepared the spaces for these tables I have been struck by the small space occupied by the malady "dyspepsia." Surely, I thought, if there be one malady which above all others should need a large space it must be dyspepsia. Yet year by year, as at the end of them I cast my eye over the tables, was I astonished to find my cases entered as "dyspepsia" to be a mere handful. How could this be? I asked myself repeatedly. "Martyrs to dyspepsia" are to be found at every street-corner, and are said to form something little less than the staple of those who drift from consultant to consultant. Moreover, I had some suspicion that I was in repute as a stomach-doctor, and was, at least, conscious of a readiness to receive such cases, and to work patiently with them. I thought also of that long-suffering organ itself and its work, of the greediness or recklessness which make it the helpless receptacle of all sorts of rubbish; I pictured to myself the contents of the viscus distended with greasy, acidulous or trashy matters, sodden in half-fermented gooseberry-juice or sour claret; and I stood amazed that it resists destruction. Yet not only does it survive, but it patiently draws order out of disorder, sweetness out of foulness. Again, as I turned to the journals of the day I read of men of lofty endowments whose lives became accursed by dyspepsia; and on other sheets I read many columns of advertised solace, from wind-pills to the pill of the American doctor, whose virtues were thus rendered at Cambridge:—

> Hæc jecur instigat, stimulos hæc renibus addit,
> Ambulat arcanas hæc taciturna vias.
> Attamen hæc eadem latebras penetralibus imas
> Mordacis buxi more modoque petit.*

Our medical journals all tell us the same story. Every large drug-house has its pepsines, its dinner-pills, its cordial bitters, testifying not only to the general, but to the medical, cry for help against the demon of dyspepsia. Finally, so confident is the public of the true form of its enemy, that one of this sort who seeks the doctor will dictate his diagnosis, declaring himself an old dyspeptic, and demanding your specific. A hundred doctors have pronounced him dyspeptic; needlessly, for who can know it better

---

* Mordacis buxi, Anglicè, "a small-tooth comb."

than himself? He will condescend, perhaps, to admit some discussion of the respective vices of his stomach and of his liver, but he is offended if you seek farther. He half tells you that if you have yet to learn what is meant by a dyspeptic or by "biliousness" you are no man for him. Patients have positively left me in dudgeon because I ventured to hint that the word "indigestion" expresses rather an inference than a fact, and pointed out that diagnosis belongs to the physician, and not to the client.

How is it, then? How are we to explain this catholic wailing over a disease which is not? this wealth of balsams for sufferings which cannot be named? Is there no distress to lull, no pain to lenify? In turning our minds to this question, let us, in the first instance, consider the meaning of the word dyspepsia, and ask ourselves if there be in it any such confusion as may account for its indefinite uses, and whether in such uses there be equivocations such as to account for the doubts to which I have referred. I believe that, if I thus follow the origin and paths of my own thoughts in the past, I shall best open out the main subject of my address. Forgive me if I thus approach the subject of visceral neurosis by an indirect method, and in the way of comment upon the affections of the stomach. It is almost trivial to remark that dyspepsia, strictly speaking, is not a disease, but a symptom; it names not the causes nor the processes of the evil, but its consequences; that, as an outcome of the affair, whatever it may be, food is not digested or is digested imperfectly or painfully. Taken in this sense, no doubt dyspepsia is an exceedingly common complaint. But if the word be used in this sense, does not its meaning become almost futile? As a consequence of various conditions, let us say, persons may find themselves unable to walk far, to walk steadily, to walk painlessly, or even to walk at all. Yet we are to use a word such as dyskinesis, and to say that whosoever comes to us walking ill, whether from general debility, from palsy, from sciatica, from muscular rheumatism, from gout, from disease of foot, ankle or knee, that he is suffering from dyskinesis? Are we to say of a patient whose forces are so diminished that he can walk only a mile, and this with fatigue, that he is suffering from atonic dyskinesis? Are we to treat all the above maladies in a chapter under the heading of dyskinesis, and in a few columns of entanglement try to deal intelligibly with such a subject in such a way? We should not be saved from the discredit of foolishness if we were to do so. Then, how are those to be spared who, using the word dyspepsia in this way, devote chapters and books of chapters to such a symptom, and raise that which is accidental and consequent to the place and name of a definite malady? Yet I have only to turn to any modern textbook to find page after page of such writing, pages to which I would willingly refer accurately were it not that I would avoid even the seeming of personal controversy with authors whose industry and thoughtfulness have done so great a work for modern medicine. If, however, we would speak wisely and strictly, shall we not seek to clear away this kind of confusion, and set out in their natural groups as distinctly as we can the several classes of symptoms now heaped together unreasonably? Thus we shall better understand the cases which come before us, and, taking a larger view of them, see better how to cure them. Perhaps, too, by this method, we shall even do away entirely with the word dyspepsia as the name of a malady, and reduce it to the position, for instance, of the word cough, which has no definite connotations, and signifies only an inconvenient and notorious symptom common to many maladies.

Sifting, then, as we must, the heap of affections attended with dyspepsia, we readily enough propose four chief groups of stomach disorders. First, those which are due to grave disease of the stomach itself, such as cancer or ulcer of the organ, dilatations of it, atrophy of its coats, and so forth. Secondly, to diseases of the stomach less grave but still local, such as the simpler catarrhs, acute and chronic gastritis. Thirdly, disorders which depend upon no visible changes in the structure of the stomach, but consist in some disorder of its work or secretions, which, in accordance with a convenient use of words, we call functional; the dyspepsia of gout may perhaps be taken as an instance of this group. Fourthly, disorders which depend, not upon any primary derangement of the tissues or secretions of the stomach, but upon some influence coming from blood or nerve, from which influence visible local changes may or may not ensue.

We easily put aside the first class of cases as quite foreign to our purpose; and, so far as they can be distinguished, we put aside the second class also. This it is easy to do in the instance, say, of chronic or acute gastritis caused by the excessive use of alcohol; or, again, in the instance of that common and transient catarrh of the stomach attended with loss of appetite, headache, coated tongue, constipation, sickliness and malaise, which is due often to cold, often to improper feeding. This affection accounts for a large number of the cases called bilious attacks, though the majority so called are probably migraine. It is easy to make some such progress as this, but thenceforth it is not easy to plan out our schedules further. We may leap over a score of doubtful cases, and, passing at once to another extreme, separate into a class all cases which are obviously neuralgic; but between these positions lie a great number of cases which are hard to classify. Is pyrosis, for instance, some local and primary affection of the coats of the stomach, or is it a neurosis? Is flatulence, are all acid risings in the stomach due to some irregular transmutation of the contents of the organ, or are some of them consistent with the course of normal, if feeble, digestion ? In the former case such derangements may fairly be called dyspepsia, for the process of digestion goes wrong, not as a detail or subordinate part of larger or systemic irregularities, but as a simple and local error in the digestive work of the stomach itself. In a word, is stomach derangement, in a given case, but an expression of some general derangement, or is it a substantial ailment, and the cause rather than the consequence of any more general perturbations ? Even this may be very difficult to estimate, or may be decided only on the study of individual instances. But a large number of cases remain which are even yet harder to interpret, cases in which dyspeptic symptoms are no doubt due to disordered work or secretion in the stomach, but in which this disorder of work or secretion may, in its turn, be due, not to any primary defect in that organ, but to some cause lying outside of its peculiar tissues, lying, for example, in the blood or in the nervous system. Omitting, as I must do, all reference to blood-changes, let us take, as an extreme case, a sudden nervous shock which may arrest digestion completely, and we may thereafter conceive of lesser degrees of nervous perturbation which may set up a continued interference with digestion, or, in other words, cause a continual dyspepsia. We may conceive, indeed, that not only in this way digestion may simply be slowed, but also the peptic secretions or processes be positively vitiated, as is the case, say, in neurotic diarrhœa. Or, to turn to a different point of view, a patient who suffers from many symptoms indicative of nervous derangement tells you that on the empty stomach, generally in the morning before breakfast, a dense yellow oily fluid gathers, and, lying there all day, would vitiate the viscus, would act as an eccentric cause of headache, and as a foul ferment upon the food in the course of digestion; so that his or her only hope of a comfortable day is to vomit or wash out these dregs at the beginning. Now, such an exudation may be due to some chronic distemper of the coats or glands of the stomach; but I incline to believe it is due rather to some perturbed innervation of a stomach otherwise healthy, seeing that the symptom is one which I have always found in neurotics, and to be curable only by treatment planned mainly upon this diagnosis. Disordered work and distempered secretions, then, may well be due, and doubtless often are due to neuroses of the stomach; and such neuroses, lying between the more localized disorders and the purer neuralgias, are difficult to classify. There is one more difficulty, though in practice a less embarrassing one—namely, that, as distempered secretion may be the effect of disordered nerve, so, reversely, some slighter catarrh or other local change, or some offending article of diet, or again, some graver local mischief, may, as peripheral irritants, set up nerve disturbances which, in certain susceptible persons, may so wax as to overshadow or conceal the primary cause of their occurrence. We know, for instance, that the touch of a bronchial attack, or of an acute pneumonia, may first reveal an asthma till that moment wholly latent—latent, it may be, till middle or even later life; or latent it might have been, like the unwept tear, for ever. But the sleeping ill, once awakened, rarely recedes altogether, and by its recurrence tends to rivet upon the sufferer the chains of habit. Thus, it may become difficult to say which is the predominant factor in the consequent group of discomforts. As a practical difficulty this is most serious in cases in which ulcer may or may not be present; and I do not hesitate to say, gentlemen, even before you, that in some of these the diagnosis between ulcer and pure gastralgia is, in certain stages, impossible. How are we, then, to succeed as ministers to the sick, if we crowd into one chapter, and almost into one point of view, the pure neuralgias of the stomach, the neuralgias awakened by local irritations within the viscus, the disordered secretions or metabolisms within it due to perturbed innervation, and the primary dyspepsias of local origin which concern the nervous system but little or not at all? Never can we succeed, I think, if we make such a confusion. We must endeavour, then, not only to separate the pure neuralgias from the dyspepsias or pains of local origin ; but, however difficult it may be in any one instance, we must endeavour further to decide, concerning mixed cases, whether the neuralgic phenomena stand in causal relation to the local disorders of function, or, contrariwise, the disorders of local function have awakened nervous reverberations. That which reason and examination fail to discern may often be revealed to us by the test of treatment. The relation of asthma to disorders of the lungs has been quoted as an illustration of these difficulties. In some cases, then, we have to deal with a pure neurosis of central origin; in others, with nervous phenomena awakened by persisting or foregone local maladies; and, unless we see with some clearness how these phenomena are related to each other, we shall fall short of a rational therapy.

In order to enter by the plainest route into the more intimate knowledge of neuroses of the stomach, let us advance from the simplest to the more complex. No cases, perhaps, will serve us better as an introduction than those in which disorders of digestion occur as a consequence of general nervous exhaustion. Unlike gastralgia, these disorders may arise in men and women of very various habit of body. No general sketch of the bodily aspect nor of the temperament of such patients can be delineated, and accordingly we find the symptoms various.

Omitting, then, persons disposed to gastralgia, we find two kinds of dyspepsia at least arising in persons overworked. In the first class, we find simply a feeble stomach, as we find enfeebled legs and an enfeebled brain. The tongue is clean—too clean; not red, but pallid. Its substance is œdematous, and its edges indented. Such a person has no appetite, and the little he eats causes a weary sense of repletion until the ingesta slowly pass off. We may call this dyspepsia, as we may call the leg-weakness dyskinesis, and so forth; but the whole man is run down —call it general dysergy, if ἀγόραια ὀνόματα are unacceptable. But now take another man, equally without marked diathesis, and equally overworked; his tongue is bulky and also indented, and is protruded slowly, spreading forth as it issues. It is thickly coated, especially towards the mid-line, where the fur is brownish, and it is brownish towards the back also. The complexion is muddy, and the countenance bears the mark of mental depression. The hand is placed fretfully upon the vertex, where there is a pain, or, if not a pain, a peculiar indescribable uneasiness, and this passes backward, surrounding the occiput. The urine deposits lithates, and the liver and stomach not only work feebly, but their functions are aberrant. He is weary and dull, and says he feels like a dead dog of a morning. A like state of tongue, and a like sluggishness and diversion of stomach, liver and colon may be seen in most cases of common apoplexy, and are clearly secondary to the troubling of the brain, though the converse is usually held, and the attack of apoplexy attributed to a foregoing upset of the digestion. Why, of the two overworked men, nervous exhaustion should produce in one a simple atony of the stomach, and in the other an aberrancy, I cannot say. I suspect the latter man has in himself some echo of gout, but yet I know how disappointing are all attempts to clean his tongue and set his stomach aright with the stock rhubarb and soda mixtures, with pepsines, with calomel, colocynth or colchicum, and how that even strong tonics, such as quinine and iron, may be prescribed with benefit, and how rest and upland-air may beat all medicines whatever, and clean the tongue in a fortnight. Dyspepsia here is a symptom, and the stomach is disordered.; but radically the state is a neurosis, vascular or other, and curable only upon this understanding.

That which I have illustrated by the examples of the stomach may be likewise illustrated by examples taken from the other viscera. Jaundice may be set up simply and directly by causes acting upon the nervous system, or it may be due to local causes only. Or, again, it may owe its origin to some admixture of the two sets of causes. So, again, in the functions of the intestines either constipation on the one hand, or diarrhœa on the other, may be due to nervous causes acting alone, to local causes acting alone, or to local changes reinforced by nervous reactions. I will only refer in this connection to one more organ out of many—to the uterus and its appendages. How intimately this organ, or this system, is associated with the nervous system is well known; but, unfortunately, the weight of our knowledge all leans one way—it leans to a curious and busy search for every local ill which may arise in the female pelvis, while blind oblivion scatters the poppy over every outer evil which in its turn might hurt the uterus; nay, more, a resolute prejudice would deny that in the woman any distress can arise which owes not its origin to these mischievous parts. *L'utérus c'est la femme* is a proverb which has received a new development in these days; for if by courtesy, rather than by conviction, woman be granted the possession of a few subsidiary organs, these, at best, have no prerogative nor any order of their own.

The uterus has its maladies of local causation, its maladies of nervous causation, and its maladies of mixed causation, as other organs have; and to assume, as is constantly assumed, that all uterine neuroses, or even all general neuroses in women, are due to coarse changes in the womb itself, is as dull as to suppose that the stomach can never be the seat of pain except it be the seat of some local affection, or that the face can never be the seat of tic-douloureux unless there be decayed teeth in the jaw. All mucous membranes, indeed, seem readily to betray nervous suffering by relaxation or changed secretion; and I make no doubt whatever that a very large number of uterine disorders which are elevated to the place and name of diseases of the uterine system are but manifestations of neurosis. All neuroses are commoner in women than in men. Facial neuralgia is commoner in them, migraine is commoner; so is gastralgia, again, and the pseudo-angina. Not only so, but in the uterus they possess one organ the more, with its own rich nervous connections, and its own chapter of added diseases and neuroses; but to say that all these maladies are due primarily to uterine vagaries, is to talk wide of all analogies. Again, some men as brave as others feel equal sums of pain far more acutely than the others; women, speaking generally, feel pain more than men do; patient as they are, they seem to have less reserve of force and less resistance, more susceptibility and resentment, and less capacity. Yet there is no standard of pain, nor of men, by which you shall say this patient is a coward and his outcry exaggerated. Men and women are variously organized in respect of resistance to pain, and their fortitude or their despair must be tested, not by their cries, but by the other features of their characters. What right have we to say that a man writhing in the pangs of a toothache is a great sufferer, while, in the same breath, we hint that a woman complaining of a pain in the abdomen is hysterical? The pain is equally invisible, equally unmeasured in the two cases, and the degree of credit to be given to the complaints is to be gauged by other probabilities. A neuralgic woman seems thus to be peculiarly unfortunate. However bitter and repeated may be her visceral neuralgias, she is either told she is hysterical or that it is all uterus. In the first place she is comparatively fortunate, for she is only slighted; in the second case she is entangled in the net of the gynæcologist, who finds her uterus, like her nose, is a little on one side, or again, like that organ, is running a little, or it is as flabby as her biceps, so that the unhappy viscus is impaled upon a stem, or perched upon a prop, or is painted with carbolic acid every week in the year except during the long vacation when the gynæcologist is grouse-shooting, or salmon-catching, or leading the fashion in the Upper Engadine. Her mind thus fastened to a more or less nasty mystery becomes newly apprehensive and physically introspective, and the morbid chains are riveted more strongly than ever. Arraign the uterus, and you fix in the woman the arrow of hypochondria, it may be for life.

Now, gentlemen, it is time we complete our reaction from this gynæcological tyranny, and that we of this College no longer permit ourselves to be snubbed by these brethren of ours, who calmly tell us, with their superior airs, that our use of such expressions as uterine neuralgia, irritable uterus, ovarian neuralgia, neurasthenia and the like, comes of a shallow sciolism, and is grounded upon the emptiness of our knowledge of uterine diagnosis. The spirit of meekness alone restrains me from throwing the same stone again, and accusing our gynæcological friends of ignorance of neuropathies and of the neurotic diathesis. That no man limited by the bounds of human intelligence can have a fine knowledge of medicine, surgery and obstetrics is true, is as true of obstetricians as of physicians, but every man can have, and must have, a good all-round knowledge of a kind which will enable him to take the main bearings of any case which may come to him. For my own part no conventions arrest me when my opinion is asked by a sufferer. The speculum and the uterine sound were invented as much for my benefit as for other people, and I feel it both my duty and goodwill to examine into every detail of a case referred to me, whether medical, surgical or pelvic. This done, I am able to judge under whose care the patient may best be placed, and the more I thus incidentally learn of surgery and gynæcology the more gladly and intelligently I recognize the extensive and dexterous attainments of those of my colleagues who work in these departments, and who can deal with such cases far better than I can. But I repeat that the main lines of the diagnosis of any malady whatsoever should be within the province and the abilities of any medical man whatsoever. What should we think of a family practitioner who made no diagnosis of a case of acute pneumonia, because his work lay chiefly with mothers, because he had not a stethoscope, and because he intended to call in a physician ? On these grounds, I say that any well-educated physician who does his duty to his cases, who does not idly turn them over to specialists, and who is armed with proper instruments of research, should know, as well as another, the main bearings of all his cases, and should claim to be heard on the diagnosis, although he may not and cannot pretend to distinguish minor differences, nor to have complete mastery of all the more refined devices of modern therapeutics. We physicians have been a feeble folk in this, we have shrugged our shoulders and submitted to gynæcological taunts in a way that may be very modest, but in a way that betrays our trust and our art. If the gynæcologists pelt us with stories of long pain and sickness uncured by medical futilities, but rapidly cured under uterine medication, we can mate their stories and check them by double the number of cases received by the physician from the sofa, the manipulations and mental abasements of narrow uterine specialism. To underrate our debt to gynæcologists, to forget the great work they have done in the past half century, were as foolish as ungracious; but, like all great movements in special fields of inquiry, it must be subject to reaction, and its results must be checked by those which have been obtained by other methods and in other directions. The wisest and most disinterested of gynæcologists now know well how lamentable have been the exaggerations, how narrow the views, and how deceptive the data of many opinions which have passed current in their school, and they are ready to declare that if medicine is not wholly to reclaim a great part of the field occupied by them, its culture must at any rate be shared with the physician. The physician has been at least as much to blame, in that he has contemptuously thrown aside many cases of genuine malady and of genuine suffering as hysteria. Even hysteria is a complaint to be treated and relieved, but the central blunder has been the stupid confusion between the hysteric and the neurotic subject. On adding up the cases in my chamber note-book for the year 1883, I find that under the three heads of neuralgia, neurosis and neurasthenia, 151 new cases were entered in that one year. What a tale of misery does this limited experience of mine indicate! Add together all the patients of all the doctors in the West Riding alone, and compute the manifold suffering! Yet I do not hesitate to say that in our vigorous northern people hysteria is far from a like extension. On the contrary, if the word be used with due care, I would go so far as to say it is rare. I remember but one case of hystero-epilepsy in the Leeds Infirmary in the last year or two, no cases of contracture, and of other hysterical affections but a few, and these often paraplegics, who relapse, and returning twice or thrice to our care swell the apparent numbers in our books.

Take a hysterical person, man or woman, in its common and, so far, proper sense; take it to mean a person of feeble purpose, of limited reason, of foolish impulse, of wanton humours, of irregular or depraved appetites, of indefinite and inconsistent complaints, seeing things as they are not, often fat and lazy, always selfish; or, to take it in less degree, one capricious, listless, wilful, attractive perhaps, yet having always the chief notes of hysteria—selfishness and feebleness of purpose; and if such persons complain of globus, of palpitation which is never perceived by the stethoscope, of sleeplessness of which the nurse has no record, of dyspepsia which does not lessen the labours of the cook, of pains which never flush the cheek; and, if such persons have or have had anæsthesia, unreal epilepsy, unreal syncope, unreal palsy, unreal cramps, then set down such a person as hysterical, but forget not, nevertheless, to cure her mind and body. Such a patient is, no doubt, a member, a degenerate member, of the neurotic family; but it is almost with indignation that I repudiate the application of the adjective to the nervous sufferer, whom we may call the neuralgic member of that group. Why, gentlemen, my neurotic patients, if I can indicate them by a name, are almost the best people in this wicked world! Rarely endowed with the capacity, endurance and profounder imagination of the greatest, they form a large number of those in the second rank who are the salt of society. Let us suppose that Mr. Galton had photographed 400 of these, inside and outside; the resulting ideal neurotic might be much as follows. His entry into your room tells of him at once. He enters with a brisk step and a quick observant eye. You see a slightly built meagre man, of sallow complexion, or, if coloured, with the colour painted high upon the cheek-bone, the cheeks and the temples are hollow, and the temporal arteries are visible under the lean skin, which often shows tanned markings deepened during attacks of pain; the hair is straight, fine and sparse upon the scalp; the features are sharp, often prominent; the lips thin and the skin dry; and some remnants of eczema may be seen about the ears or chin. The tongue is protruded and retired quickly, and is generally narrow and pointed; it is rarely indented, and its tip, even when the health is best, does not cease to be red: there is often a light silvery coating upon the dorsum and mid-line. The bodily frame is lightly and often finely built, the bony fingers and wrists and the visible sinews and radials betraying the absence of fat. Here and there, in later life, a knotty knuckle may tell of gouty parentage. The pulse when most tranquil, usually ranges between 70 and 80, and accelerates on the least excitement. The clavicles and ribs, in like manner, are prominent, and the heart's apex may be seen to beat sharply before the eye: its systole to the ear is likewise short and sharp, and the second sound very audible over a wide area. The limbs are small, but often very sinewy; such persons are as active as birds, and the absence of fat in their muscles often gives to these, in states of health, the quality of hardness under the hand. Their conversation, again, is lively and voluble, often keen and brilliant, but impressionable rather than imaginative; you may generally notice in them, too, some little blinking, twitching or tattooing trick which quickens as thoughts and words come faster. Usually, such a patient does not readily come to you; he is brought, half reluctant, by his wife or friend; he says, apologetically, he is an old dyspeptic, and you can do him no good. He has visited all the springs and half the doctors in Europe, and he lays a bundle of old prescriptions upon your desk. Once agate, however, his story will be a long and minute one, but never maundering, wandering nor whining. His companions will tell you that he is subject to great fluctuations of the animal spirits—gay, even fascinating, in society; brisk, orderly and thorough in business, but at home dejected or fretful. He is a small eater, a light sleeper and a worn worker. These persons are the heirs of every true neurosis, from insanity to toothache; and on the whole, when we consider the infinite perturbations of intermarriage, it is surprising how true they run, or how clearly you may detect the neurotic strain in mixed descendants. Of their visceral neuroses, I shall have to speak hereafter, and would only say now that in both sexes of them migraine, stomach-ache and windy colic are frequent and eminent, and receive the name of dyspepsia; and in the women are added to these uterine and ovarian neuralgias and hyperæsthesias. To call these suffering women of the neurotic type hysterical is to confuse all due acceptance of names, and, what is worse still, it is to confuse the real relations of things. The neurotic woman is sensitive, zealous, managing, self-forgetful, wearing herself for others; the hysteric, whether languid or impulsive, is purposeless, introspective and selfish. In the one is defect of endurance, but in the other defect of the higher gifts and dominion of mind.*

Now, if we turn our eyes upon the flock of women who lie under the wand of the gynæcologist, we shall find it so largely composed of the neurotic and hysteric, that we may say in our haste the uterus has no substantial diseases; that its affections are all neurotic, or so far reinforced by neurosis as to depend for their cure mainly upon neuropathic medicine. Herein we in our turn should be to blame. Many a woman, otherwise robust enough, and many a woman whose weakness may lie not in her nervous system, suffers from uterine disorder, from painful uterine

* One very kind critic of my lectures has suspected that any overworked man may be reduced to the state of my neurotic. This is not so; *nascitur non fit*. Of some of the consequences of mere atony in fagged men not of neurotic habit, I have spoken already on pp. 12, 13.

states, nay, even from distant sympathetic pains also, which come of mischief wholly local, or of mischief reinforced by diatheses other than the neurotic. Making, however, the utmost allowance for all these, I contend that a vast number of such sufferers lie under the scourge of neurosis, and that their uterine and ovarian disorders are either wholly neurotic, or, as I have said, so reinforced by neurosis as to depend chiefly or wholly upon general medicine.

Let us take as an instance a young lady coming of a family in which great mental gifts had thrown into relief the many eccentricities and humours which accompanied them; a family, too, of which no household had been free from nervous disease. She possessed the gifts and the attractions of the neurotic diathesis, and laboured under its defects. It is possible also that she was in some degree under the stress of what Anstie called the unconscious sexual impulse. She was restless, excitable and suffering. Her pains were mostly pelvic and abdominal. She never put her feet to the ground, partly because it intensified her pain, partly because she had been forbidden to do so. She had lain on her back for months. Pessaries had often been introduced, but being intolerable to her were withdrawn. Her periods were agonizingly painful for the first two days, and were profuse; and she had constant leucorrhœa. Her appetite was almost gone, her stomach queasy, her frame emaciated; but she was full of courage, unselfish, and would have scorned the wiles and exacting whims of hysteria. Her womb had been incessantly under specular and other examination for a year or two, and, like nearly all such patients, she had uterus on the brain. I found the vagina tender, and the womb exquisitely so; its substance was soft, and its attachments lax. Its position, therefore, was somewhat backwards and downwards. Acute suffering was caused in the upper hypogastrium when the fundus of the uterus was pressed upon per rectum. The rectum was full of fæces. By the speculum I noted there was both uterine and vaginal catarrh, and that the os uteri was excoriated—in the state, that is, of the upper lip of a scrofulous and snivelling little boy. My most difficult task was to win my patient over to the belief that her disease was not entirely uterine, but mainly neuralgic; this once accomplished, our progress, though slow, was sure. I declined to initiate any treatment whatever until she would get her feet to the ground, and thenceforth cautiously regain the use of her legs. Meanwhile, I declined to "cure the ulceration of the womb" for the twentieth time, but made her content with rectal and vaginal astringent douches, first hot and afterwards cold. As soon as she could walk we perched her upon horseback. She was treated with the phosphide and valerianate of zinc, with bromide of ammonium, iron, quinine and like remedies, with occasional sedative suppositories. In six months, I found the uterus more compact, the ligaments braced, and the os clean and sound ; the leucorrhœa had ceased, and all the parts could be handled without pain. Menstruation was still painful, but less so than formerly, and there was some menorrhagia. She was mixing, however, in general society, could ride gently to hounds, had regained appetite and looks, and, although I then lost sight of her, I have every reason to suppose she is as well as she is ever likely to become.

Now, gentlemen, is not this case one which in their degrees could be multiplied a hundredfold from our case-books or our memories ? Yet these are they who form a great part of the women who are caged up in London back drawing-rooms and visited almost daily for uterine disease, their brave and active spirits broken under a false belief in the presence of a secret and overmastering local malady, and the best years of their lives honoured only by a distressful victory over pain.

The case I have described was selected by me because it was not one of mere irritable uterus without apparent disorder. Irritable uterus, in spite of the denial given to it in high places, is a genuine malady. It corresponds to the hyperæsthesia of the stomach which is found in the same diathesis, and which often simulates ulcer of the stomach.

But I pass over all these as neurotic enough, and I assert that, in such neurotic subjects, uterine laxities, moderate displacements and catarrhs owe their continuance, and often their very initiation, to an atonic state of body, and to a special instability of nerve-endowment, which may show themselves in failing function, and soon after in local trophic changes and perverted secretions. Such changes of function and such settlements of perverted action are often, no doubt, called to this spot or the other by some local deviation from the normal, as a consumption may take its beginning from some trivial and forgotten catarrh; but the essence of the malady is not there, and to try to cure such a malady by local means is as wise as to try to cure a syphilis by antiseptic dressing of its ulcers. Such subsidiary means are often needed, often indeed necessary; but in cases like those under discussion should be used as little as possible, because of the tendency of such methods to arouse and perpetuate a morbid possession of mind in the woman. All this our more robust, more clear-sighted and more candid gynæcologists know well enough; in the rest the fault may lie rather with modern fashion than with themselves. Looking only to the uterine organs, their reason bounded by the confines of the pelvis, they attempt to stem the tides of general and diathetic maladies with little Partington-mops of cotton wool on the ends of little sticks. That many of the cases we have discussed need a judicious combination of local with general treatment is true, but in most of them the patient and the doctor are fascinated by the local phenomena, while Nature herself is performing on a far larger scale. If we are to cure disease, we must be able to fly with her and to run with her, as well as to creep with her. In my later chapters I shall recall the truth which should be ever before us—that the fundamental difficulty in all neurotics, not hysterics, is their nutrition. More fresh air without expenditure of the slender store of strength, the permeation of their starved tissues with the fat that they themselves so often loathe in their food—these two reforms accomplished, all their organs will take on a more generous and a more vigorous life, all their tissues will brace and cleanse themselves from a purer and richer fountain of blood, and force will be stored up and energy developed, wherein before were dilapidation and sterility. As a shrewd old Yorkshire doctor once said to me, "It's no use, my lad, putting the hands right upon the clock-face if you haven't cleaned the works."

Gentlemen, we are all one-sided; I speak to-day from my own one-sidedness, and my convictions are upon the side of cleaning and repairing the works. Sometimes, no doubt, when the general state of the health is restored, some local trouble set up originally by the constitutional state—be it gout, scrofula, or what not—smoulders on, forgotten as it were, after the general malady is cured. For such local trouble diathetic treatment no longer avails; it must be wiped out by some local alterative. But, on the other hand, to contend daily with local troubles which daily are regenerated by some vicious habit of the whole system, is to roll up daily the shameless stone of Sisyphus.

LECTURE II.

Mr. President and Gentlemen,—We all know the story of the resentful lion who had so often been painted by the man, and who longed for the time to come when the man in his turn should be painted by the lion. Is it because all medical books have been written by men that gastralgia, one of the sharpest arrows in the armoury of pain, is in many of them dismissed with a few words as a malady of " hysterical women"—of women, that is, whose sufferings are due in part to effeminate habits and constitution, and in part to a kind of fanaticism which prompts them to cherish or to imagine pain, or to make a bitter cry about small matters ? We all ought to know that gastralgia is common enough in man also, though certain writers appear to deny that this or any other visceral neurosis can exist where no uterus is; and in my first lecture I have endeavoured to formulate a distinction between what I may call the upper class of neurotics and their degenerate relations the hysterics; asserting therein that in hysterics pain is not a more but a less common inheritance, and that to neurotics it is chiefly given to suffer pains and the renewal of pain, and to bear all with singular fortitude and spirit. As such a zealous neurotic the Sompnoure posed when he described himself thus :—

> I am a man of litel sustenance;
> My spirit hath his fostring in the Bible;
> My body is ay so redy and so penible
> To waken, that my stomak is destroied.

For of all the neuroses of the stomach gastralgia is the chief; but around it gather, in more or less close association, flatulence, vomiting, hyperæsthesia, and miseries such as distension, sinkings, cravings or loathings of food, and these may exist in various associations, or may exist singly. As in angina pectoris and the pseudo-angina, so in gastralgia, the spinal nerves may be included in the paroxysms, or may take even a chief part in them, the visceral and overlying spinal nerves being grouped in function and in suffering together. In order to study this class of cases more carefully, I have gone over the case-books of my chamber consultations for ten years past—namely, from the year 1874 to the year 1883 inclusive. From them I have extracted 139 cases of gastric and abdominal neuralgias. These cases remain after eliminating all that are doubtful or comparatively trivial in degree; all cases attended with important uterine disorder, or complicated with substantial defect in any organ whatsoever, or with disease; all hysterical cases, and all cases of fretful or whimsical persons prone to expand or to exaggerate their symptoms. The selected cases, then, are the cases of persons of whom I have some tolerable notes, who were straightforward and truthful, and who are, or were, the subjects of neurotic or neuralgic affections lying under the diaphragm and above the pelvis. Now, the great majority of these cases are gastralgics, the region of the stomach being by very far the commonest seat of abdominal neuralgia. Of the gastralgics the majority are women, in the ratio of two to one. In men and women gastralgia is by no means confined to middle life, as Leared supposed, but is found at all ages, from fourteen to sixty, being most commom between twenty and forty-five. Like migraine, it tends to die out in middle life; and, like migraine again, it is attended less and less with vomiting as age increases. Gastralgia comes on earlier in women than men, apparently by some ten years. A gastralgic man is rare before the twenties; girls often begin in their teens. The earlier maturity of the girl sufficiently explains this difference, which is also remarkable, though I think in less degree, in migraine. Migraine may, however, begin at very early ages in either sex. Of associated affections migraine is by far the commonest; and if we associate migraine and neuralgia of the head and face together, the number of cases in which these are found with gastralgia is very great, seeming to be fully 80 per cent. Fortunately these affections do not often coincide in time, but rather alternate variably one with the other. A large number of gastralgics, then, include migraine or neuralgia of the head and face in their life histories. Another disorder commonly associated with gastralgia is asthma, an observation which scarcely needs either explanation or much asseveration. Neurotic vomiting may accompany, or rather alternate with asthma, as the gastralgia may; and so again may voluminous flatulence, borborygmi, and other such neuroses. Like asthma, gastralgia is not unfrequently nocturnal in its recurrence, and is likewise often aroused by improper food, or by the contact of any food whatever. Hence the persistent confusion of gastralgia with dyspepsia. Asthma and gastric neuroses, moreover, if not concurrent in the individual, are often concurrent in families; one member suffering from the latter, and his brother, sister, or other relative from the former. With both these vagus neuroses run the cardiac neuroses, including true angina. I am now attending a gentleman with Mr. Jessop, who suffered terribly from paroxysmal gastralgia in past years, and in whom we have now to deal with the true angina. In gastralgics occur violent attacks of palpitation, or, more characteristically, attacks of slow or intermittent pulse, as in the following case.

Mrs. ——, aged twenty-four (No. 371, 1875), suffers from marked gastralgia. She has one child which she is nursing. She consults me now for pain in the region of the heart of a dull prolonged kind, ending in a sense of exhaustion. The pain may strike to the back and under the left shoulder-blade.

At such times the heart becomes very slow (48 to 50), intermittent and irregular. Curiously enough such attacks are brought on at once by putting the child to the breast, even before much milk is withdrawn. This sequence is analogous with the sequence of gastralgia after the ingestion of food, the neuralgia being awakened by a peripheral irritant. She soon recovered on appropriate treatment; but in 1881, after some menorrhagia, the same thing recurred. She again recovered on like treatment. She had had one child between these dates without any disorder of health.

In No. 371 appeared moderately a radiation of pain upon the spinal nerves; this, in other cases, may extend or intensify itself to a degree of pseudo-angina which may even be indistinguishable from real angina.

No. 283, 1878, male, aged thirty-seven. Of nervous temperament and a zealous worker; he is liable more frequently to violent gastralgia, especially after worry and excitement. Less frequently he has "heart-spasms." In these he has a sense of swelling at the heart, and pain and numbness down the left arm. At the same time the pulse becomes "much enfeebled," and is said to be slowed down as much as twenty in the minute, or may even become imperceptible (this from his own doctor). These attacks have come occasionally for many years. Physical exertion does not call them forth. All organic factors seemed normal.

Cases like these may be culled from every year's entries. In one lady, aged thirty-three (No. 261, 1881), wholly free from menorrhagia or leucorrhœa, an intense aching, with a sense of extreme exhaustion and sinking, would come on at the heart; it recurred then daily, and was always periodic, appearing about 4 or 5 P.M. She was compelled at such times to lie down and "press the heart," turning over on the left side towards the prone position. These attacks were encouraged especially by any abbreviation of sleep or by fatigue. She herself is subject to gastralgia without dyspepsia, and her mother was mildly epileptic.

Dr. Ramsay, of York, has often consulted me concerning Mrs. ——, aged forty-three, who more than once, with miscarriage, had lost blood heavily. She is very subject also to menorrhagia, which is scarcely kept in check by close attention to local and general remedies. She is, however, wholly free from any persistent affection of the womb or of other organs. She suffers occasionally from gastralgia, but her great distress consists in attacks of stabbing pain, as if the apex of the heart were pierced by an arrow; the pain thence extends with intense cardiac oppression down the left arm to the fingers, causing a numbness there. The pulse falls from an average of 70 to 60, but is not otherwise much embarrassed. She has the faintness and dread, and, in a word, her symptoms are not in themselves to be distinguished from angina pectoris.

The same difficulty of immediate diagnosis is found in the very similar case of Mrs. ——, a patient of Mr. Lee of Dewsbury. In her too is a persistent disposition to menorrhagia, and also to utero-vaginal catarrh. In these and like cases diagnosis depends less upon the character of the attacks themselves, and more upon the other records. The attacks are commonly due to such losses as menorrhagia or severe leucorrhœa or prolonged mental distress; they are not directly produced by bodily exertion, but are often caused indirectly by the resulting fatigue. In some of my cases, again, the pains are more extensive than is usual in angina. In some the pain runs down both arms alike, or runs down both the arm and the leg of the left side, or again springs upwards from the cervical plexus to the same side of the face and head. Valleix' points may be detected in the course of the spinal nerves, but in many cases of pseudo-angina they are not definitely to be distinguished. In women we may rely on the lesser frequency of true angina in them, though the pseudo-angina is often met with in men. About 20 per cent. of my cases of pseudo-angina were in men. In one gentleman, of neurotic habit and ancestry, and whose family affairs were laden with sorrow, such attacks occurred, involving seriously the action of the heart itself, and producing the dread, the ashen countenance, and other marked signs of true angina. As his luck in many ways turned for the better his attacks diminished, and I believe he has now been free from them for nearly ten years. A favourable opinion in his case was founded upon his age (thirty to thirty-six), his diathesis, his unhappy circumstances, and his organic soundness. The tendency in all these cases to retardation of the pulse brings them, apparently, into the group of vagus neuroses with asthma and gastralgia; nor will intercostal and cervico-brachial neuralgias cover all their phenomena, such as the dread, the pulse-changes, and the peculiar pains which such patients refer instinctively to the heart (as in Case 261, 1881, *supr. cit.*)

It is perhaps impossible duly to attribute to each nerve or system of nerves its own share in these phenomena; even the essay by Professor Eulenburg seems to me to leave the matter pretty much as any intelligent observer may well have guessed it for himself. Still I think it seems probable that the phenomena of pain and spasm are due to the pneumogastric nerve and the intercostal and spinal nerves associated with it, while such other sufferings as faintness, sinkings, cravings, distensions, palpitations, abdominal pulsations, flatulence, retching and vomiting, pyrosis, diarrhœa, lithatic or watery urine, insomnia, and so forth, may be attributed to the sympathetic alliances. Pulsations of the abdominal aorta in neurotics, whether men or women, may be so definite, may simulate aneurism so closely, and do so often appear after some history of effort or strain, that diagnosis upon direct observation may be as difficult as in the case of pseudo-angina, and may as therein derive most of its force from the history of the patient and of the family disposition. Avoiding digression, however, and returning to the association of pain in the spinal nerves with pain in the viscera, we observe that cervico-brachial and intercostal pains are found with gastralgia as well as with the pseudo-angina. Out of many cases of the kind I may refer to Miss —— (No. 504, 1883), who suffers from gastralgia of the ordinary type, but severely. The attacks are nocturnal. The pain runs down the left arm, until the arm feels quite benumbed and useless. She has faint feelings, but no anginiform distress; her catamenia are regular and normal, and she has no leucorrhœa. Her family history is very neuralgic. She bore arsenic badly, but was quite cured by a careful use of it.

Mrs. ——, aged thirty (No. 403, 1881), had visceral neuralgia ten years ago, and now for two years past. She has ordinary epigastric gastralgia, unconnected with digestion; and her attacks are usually nocturnal. The pain runs into the left submammary and axillary regions, and into the left arm. At other times she had suffered from recurrent enteralgia, similar in character to the former, but then running down into the groin. Her uterus and ovaries are normal, and her catamenia regular. She has no leucorrhœa. At the beginning of her attacks she passes urine frequently—say six times in an hour; then, as the attack subsides, the bladder likewise goes to rest. She came to me on account of frontal and parietal cephalalgia of fourteen days' duration, during which time she had had no visceral pain.

Mrs. —— (No. 38, 1877), aged thirty, has had migraine and cephalalgia for years. Her sister is a pronounced gastralgic. For two years she has been subject to attacks, "pinning the shoulders tight," and girdling her with pain round the waist. The pain almost stops the breathing, and so continues for two hours. She then vomits, and the pain departs. She never had jaundice. She looks worn out with pain, but all the organs and functions seem normal.

Mr. ——, a cashier (No. 86, 1876), aged twenty-eight, had pains, as in the preceding case; they were periodical in recurrence. But in his case the attacks culminated not in vomiting but in diarrhœa. He recovered completely on tonics, with rest and change of air. He had a neurotic history, personal and inherited.

Miss ——, aged thirty (No. 51, 1876), presented symptoms almost identical with those of No. 38, 1877, including the critical vomiting. She was also subject to morning vomiting, attended by frequent micturition. She used to have headache with the vomiting but not for some time past, though she has severe temporal neuralgia occasionally.

Another very curious symptom which occasionally occurs with gastralgia is yawning. I have seen this in many cases, and may quote P. T. P——, a grocer, aged thirty-three (No. 571, 1882), who suffered from gastralgia in the usual way, but with an incessant and irresistible tendency to yawn.

Vomiting is a symptom which, in some cases of gastralgia, takes a very prominent place, but in others is never seen. I have already drawn a parallel between the vomiting of cephalalgia and the vomiting of gastralgia. In either case the vomiting may stand alone—may stand without headache, or, again, without stomach-ache. In such cases it may only be known from the case history whether the vomiting belong to the one order of things or to the other; but the vomiting allied to migraine is more commonly in the morning, is more periodic, and recurs at longer intervals; the vomiting allied to gastralgia is more dependent upon the ingestion of food, is irregular in recurrence, and may, indeed, be daily or almost incessant. As an example of the former, of migraine without headache that is, let us take the following case.

Miss ——, aged twenty-seven (No. 14, 1880), comes of a neurotic family, and she herself had had migraine for nine years. Then the headaches gradually wore off, the vomiting part of the attacks only remaining. This was not very violent, but returned at intervals and would last for about two days, during which time she had to keep as quiet as if she had a sick headache. Mental work or excitement most readily recalled the attacks. Her uterine functions and those of the digestion and bowels were all normal. This patient was a lady of attractions both personal and mental, and her case presented many interesting features. Her vivid perceptions, alertness, incessant and beneficent activity, and utter unselfishness marked her as a type of the more highly endowed neurotic, and I took much interest in her case. A good deal of amendment was obtained by pointing out to her the truest economy of strength and work, and by insisting upon complete rest at times of malaise.

Mrs. —— (No. 22, 1880), presented similar symptoms. She used to have full migraine; but from it the headache gradually disappeared, and left only intense and prostrating vomiting which would recur much as the full migraine used to do, and, like it, would last for five or six hours and then vanish. In her no functional derangement was present, and she derived great benefit from a mixture of bromides and hypophosphites. The same combination was also very valuable in the preceding case (No. 14).

In such cases the vomiting, though a visceral neurosis, would seem to be of cerebral origin.

Let us now turn to cases in which vomiting seems rather to be a limb or remnant of gastralgia. The first is such a complex case as we might well expect to find. No. 44, 1875, male, aged fifty, for "all his life" has been liable to sick headaches. But these would often alternate with gastralgic attacks, which were also attended with vomitings. He is of very nervous temperament, has many cares, and no holidays. Facial neuralgia is a common ailment in him; but, for two years past, his trouble has mainly been vomiting in the morning, a little phlegm only being ejected. Day after day he finds himself unable to retain food until dinner-time. Now, this seems to have been gastralgic vomiting; and he was cured by a course of bromide of ammonium with hydrocyanic acid, followed by a course of Easton's syrup, with much general good advice. Two years later he had had no return of the vomiting nor of the gastralgia. I should add that he was a total abstainer of seventeen years' standing, and is so still. Now it is somewhat remarkable that, although vomiting is common in gastralgia and is also common in migraine, and although migraine and gastralgia may commonly arise in the same patient, yet migraine and gastralgia seldom or never occur together. I suspect, therefore, that the vomiting of migraine is not identical in mode with that of gastralgia.

Although, as I have said before, vomiting is sometimes a climax-symptom in gastralgia, as in migraine, yet in the former case it is, perhaps, reflex; in the latter it is probably allied to the vomiting of encephalic tumour or of concussion. Take again, for instance, No. 25, 1878, male, aged twenty-three. At times of a morning he awakes in low spirits, and feels apprehensive and miserable. His appetite has vanished, nausea comes on, he vomits, and, after free vomiting of nothing, is all right. Now, this young man was quite cured by guarana. I have no history of frank migraine in him, but, like the former cases, I believe it to be migraine *sine cephalalgia*.

But Mrs. ——, aged thirty-five (No. 128, 1878), carries us, on the other hand, towards the stomach. She has leucorrhœa, a soft uterus, and flabby muscles everywhere, her heart-sounds being short and weak. She has no uterine catarrh. Her catamenia are disposed to anticipate, without being any the less in quantity. She is subject at times to intense neuralgia at the supra-orbital notches and in the orbits, "poking her eyes out." Now every morning before rising she has nausea which proceeds to vomiting. This is especially worse when she is much below par. There is also a great expulsion of inoffensive wind, and clear water runs from her mouth. At the same time there is pain between the shoulders. Her appetite is never good. She is very temperate in the use of alcohol. The family history is phthisical.

Here we clearly have to do with the stomach, and I regard pyrosis as being for the most part a pure neurosis, probably vaso-motor; for the following case of Mrs. ——, aged forty-one (No. 503, 1881), takes us a step further. Seventeen years ago she fell ill, as at present, but on the whole she had enjoyed fair health in the interval. Her attacks are as follows. A sensation of languor and exhaustion is followed by intense pain at the stomach, and vomiting soon sets in. She brings up clear water, and clear water only. If food comes, it is by chance, for the attacks have no connection with diet, nor with mealtimes. The attacks may recur even as often as three or four times in twenty-four hours, and with them she has also a pain in the back, "screwing her to death." But not unfrequently she has the vomiting and ejection of water only, unattended then with either pain. She never has any nausea. Her uterine and other functions are normal, and the attacks do not favour the catamenial periods. All commemorative facts, personal and relating to family, have a neurotic quality. This lady recovered her health at the time under the use of arsenic and like remedies, and remained well until a little while ago, when she called upon me, complaining of some return of the vomiting, but without any gastralgia. The stomach-ache, she said, "used to be terrible."

Here, then, we have in close association, and evidently of one nature, gastralgia, vomiting and pyrosis or the ejection of water. In the similar case which preceded it, we had the ejection of volumes of wind also. This, I say, would lead us to believe that pyrosis, for the most part, is a gastric neurosis; and if we are then led to inquire into the personal and family histories of the subjects of pyrosis, we shall find that, as a general rule, these persons present evidence of belonging to the neurotic class.* Indeed, the regurgitation of food itself is by no means always due to an error of digestion; it occurs in many neurotic persons as a purely reflex neurosis during normal digestion.

Henry B——, aged twenty-five (No. 631, 1881), tells me that for some time he has been unable to prevent the regurgitation of food after all his meals, which is very annoying and unpleasant to him. He is not conscious of any dyspepsia whatever, nor can I find any evidence of it. His tongue is clean, his bowels are regular, and his urine of normal quality. He called upon me because of a severe attack of neuralgia in the infra-orbital and dental branches of the right side of the face. No local source of irritation could be found. The remarkable thing is that on that day and thenceforth the regurgitation wholly ceased, and under the use of quinine the

* One of my audience told me, after my lecture, that in himself violent supra-orbital neuralgia is wont to resolve itself with a copious watery defluxion from the nostril of the same side, and I believe the experience is not uncommon. This gentleman told me, by the way, that he could generally stop his pain by a copious hot water enema.

facial pain ceased also. He had no return of either disorder during my observation of him.

The grocer, P. T. P—— (No. 571, 1882), who yawned so much, had also this vexatious regurgitation of mouthfuls of food after his meals. The symptom is doubtless due to an irritability of the muscular walls of the stomach setting up an excessive, and it may be an inverse, contraction of the viscus under the stimulus of digestion, and is a phenomenon of the same order as painful hyperæsthesia of the inner surface of the organ, of which I shall have to speak at length. Before passing to that symptom, however, I must leave regurgitation, to consider another form of vomiting—vomiting, that is, not on the empty but on the full stomach. This terrible malady is described commonly enough, yet it seems to me that it is but ill discriminated, being mixed up with hysterical vomiting on the one hand, or handed over to the gynæcologists on the other. Uterine vomiting, on the one side of it, is to be known by due investigation of the case, and is curable by uterine medicine or surgery. Hysterical vomiting, on the other side of it, is not so easy to separate from the malady I have in hand, unless there be, as for the most part there is, the broad difference that hysterical emesis is consistent with fatness; gastralgic emesis never is, but leads straitly to emaciation. In many cases the vomiting is preceded by pain, more or less severe, and in these ulcer is often suspected, but in others there is no pain whatever. In both cases—in the case of those with pain and those without—the cause is cognate, if not identical. Let me instance a case from Dr. Salter's work on Asthma (ed. 1868, p. 256) —one adduced by him as an instance of "perverted innervation of the pneumogastric nerve." Condensed to the utmost the story is this.

A little girl, aged eight, rejected food of any kind the moment it was swallowed. There was no pain, no tenderness, no feeling of sickness at any other time. After vomiting had emptied the stomach she would be quite well. The same thing occurred with any little matter, such as blackberries, that she might eat between meals. She fell into weakness and emaciation, all remedies being useless. Some years afterwards Dr. Salter made inquiries about her, and found that the vomiting had been supplanted by spasmodic asthma. The two maladies had, indeed, alternated with each other at times, the vomiting coming on when the asthma was better.

It is enough to make one's blood boil to hear of a poor girl, wasted to a shadow by such a distress, treated, almost with gibes, as hysterical, when she is the subject of a malady as real as the doctor's own toothache, which, in its turn, might vanish, by the way, the moment he should find himself on the dentist's doorstep. I have a case in my mind which was recorded as hysteria by the late Mr. Skey, in the *Medical Times and Gazette*, 1866, vol. ii.* In this case intense suffering followed the ingestion of food. The poor girl had foolishly yielded to the dread of food, and was in a pitiable state of starvation and distress; and yet, because Mr. Skey succeeded, as an entire stranger might well do, in diverting the sufferer by bright conversation on pleasant subjects, he neatly labels her as hysterical. I shall return to these cases presently. Hysteria may simulate gastralgic vomiting, as it may simulate asthma, but we should do ill for that reason to treat all asthmatic girls as hysterical, because their dyspnœa is neurotic. Happily, these cases of obstinate gastralgic vomiting, for so I will call them, are not very common, if we carefully separate them from hysterical imitations of them, from gastric catarrh, from alcoholism, from uterine vomitings, and so forth. Still there are too many of them, small as the class may be, and we may properly call the malady gastralgic vomiting, seeing that the mode of them is the same as of the cases wherein pain is also a factor. In some gastralgics, for instance, the two symptoms exist but are not synchronous—the vomiting and the gastralgia replacing each other, as the vomiting and the asthma replaced each other in Dr. Salter's case above cited; or the vomiting may occur at one hour or under one set of circumstances, and

---

* For this reference and much help of the kind I am indebted to the invaluable labours of Dr. Neale, the compiler of the Digest.

the pain at another time and under other circumstances. For instance, in one of my patients, painless vomiting followed the meals, and at a later date gastralgia of the ordinary type came on in nocturnal seizures, after the manner of its cousin asthma. I find another case, under No. 112, 1881. Miss ——, aged twenty. At sixteen years of age she was subject to facial neuralgia, then to pain in the left chest and down the left arm. This left her in eight weeks, and she remained well for some months. Then gastralgia appeared. At first, this arose immediately on eating, or within ten or fifteen minutes thereafter. Then this pain vanished, and she became subject to gastralgia on the empty stomach only. This in its turn disappeared, and vomiting set in. The vomiting had gained upon her, and now she vomits all her food. It comes up in a mass as eaten. She never brought up blood nor mucus. She is emaciated, but all functions and organs seem normal. There is no evidence of hysteria. She made a good recovery on arsenic.

We are thus led to the cases which simulate gastric ulcer; and in which, as I have said, diagnosis, often difficult, is sometimes impossible. No doubt I have at times mistaken a case of ulcer for gastralgia, and at other times have assumed the existence of ulcer when gastralgia and gastralgic vomiting alone were present.

Such cases, if not familiarly known to the profession, are, at any rate, so far known to the closer observers as to make it unnecessary for me to give the long case-histories which such patients afford. Taken shortly, we find this form of gastralgia most frequently in young women; and it is marked by the sequence of ingestion of food, severe pain in the stomach, vomiting and relief. The pain may occur only on taking food, and vomiting may give speedy relief. It is not an uncommon thing, we know, for young women to produce blood in irregular ways by the mouth, sometimes repeatedly or in quantity, even when all parts are free from disease. If there be any appearance of blood, the diagnosis of ulcer becomes almost inevitable. Indeed, blood or no blood, it is often hard to come to an opinion which shall be confident enough to justify us in withholding food on the one hand, or, on the other hand, in pressing stomach-nutrition gently but firmly forward. But according to the correctness of our judgment must our treatment be helpful or positively mischievous. Probabilities do not aid us much, for I think the ulcerous and the pseudo-ulcerous cases are about equally common. The catamenia may be absent or present in either case, and I do not agree with a writer in Quain's Dictionary that gastric neuroses in women especially frequent the catamenial periods. In so far as these act as general depressants of vitality they may open the way to an attack, as may any other exhaustion; but I do not find among the genuine gastralgics a consensus of cases showing any more direct or any well marked association of the kind. Emaciation is, perhaps, more usual in the neurotic than in the ulcerous disease, but this cannot go for much, nor can chlorotic anæmia, which may be seen in either form. Pain in the back again is as common in gastralgia as in ulcer. In these doubtful cases we must try to reach safe ground by a careful review of the patient's life-history, diathesis and associated symptoms, and we can often attain this position by the discovery of neurotic antecedents, and of some previous irregularities in the now associated symptoms. We shall find, perhaps, that vomiting and pain have been dissociated in the past; that the pain has been nocturnal, or otherwise periodic; that there have been voluminous flatulence, cravings, sinkings and so forth. With such commemorative symptoms, both as regards stomach and other parts, we may sometimes recognize gastralgia clearly enough, even in spite of the occasional appearance of some ejected blood. Hyperæsthesia of the stomach is not, of course, always, nor even generally, associated with vomiting. Many persons have slighter degrees of such hyperæsthesia, and suffer some discomfort during digestion, either because digestion is slow and defective, or because the stomach resents even a normal digestive process. When such discomfort becomes an ache, the patient calls himself a dyspeptic; when the ache rises into an agony we have a very grave malady to deal with, even if there be no vomiting nor tendency to vomit. This state of stomach much resembles that of the uterus in the irritable uterus of Gooch, and of the spine in the irritable spine of Teale and the Griffins. It cannot be doubted, I think, that these three maladies, coincident as they often may be with hysteria, are nevertheless essentially independent of it. Irritable spine is found frequently, no doubt, in the plump, listless and egoistical victim of hysteria ; but it occurs likewise in the higher class of neurotics. I am attending such a case of spinal irritation at present, with Mr. Richardson in Leeds. The young lady is not only cheerful, alert, unselfish and worn in flesh, but she is full of will and energy in her efforts to carry out any troublesome plan of treatment we may suggest. I take her complaints as literally as I do the complaint of any other sensible and temperate person who says that he is suffering from some unseen distress.*

Irritable uterus, again, in spite of the mechanical school, is a genuine neuralgia in many persons, as it may be an affected malady in others. Irritable stomach is a grim reality in no less a number, though hyperæsthesias and anæsthesias, as functional disorders, do, no doubt, belong also to the hysterical; it may, indeed, be hard to say in a given case where active neurosis ends, and where a brooding persua-

* I cannot forbear to refer in this place to the admirable service Dr. Handfield Jones has done in warning us against the delusive phrase of "hysterical pain." Dr. Kent Spender offers the same testimony in his work on the "Relief of Pain," p. 65.

sion of pain begins. We have types intermediate between the sleek religious woman who knits antimacassars and lives upon her friends, and the hollow-eyed, fiercely active wan woman, who, labouring ever for others, tries to hide the wolf within her, even from herself. Before Laségue and Sir William Gull published their cases of anorexia nervosa I had myself collected six cases, with a view to publication, which thereon became unnecessary. For these cases I would deprecate the stereotyped use of the name anorexia hysterica. In some of them there is great pain on the ingestion of food; in others there is no pain, but simply a distaste for food. This distaste has arisen in most of them, however, from some discomfort which at one time accompanied digestion—a discomfort which may have taken its rise either in dyspeptic or neurotic causes. Sometimes the distaste has taken its origin in a mere shirking of food —in the fear of growing stout, or in a nobler avoidance of self-indulgence. However established, the distaste, as we know, often becomes invincible, even when eating is unattended with gastric hyperæsthesia. If I were to judge by my own original cases I should fail to see how the word hysterical could usefully be applied to any of them; in three of them the word would be absurdly out of place unless we are to apply it to any nervous malady in any woman whatever. This anorexia nervosa is no whimsical malady, no inconsistent or irregular indisposition,

but is a definite complex, consisting in part of objective symptoms. With more constancy than is commonly found in disease, we find the malady as follows. It occurs in young women, and generally in persons who inherit a decided neurotic taint. Five objective symptoms are always present—namely, amenorrhœa, constipation, subnormal temperature, slow pulse and cold extremities. To these we may add a degree of emaciation proportionate to that of the malady, and a tendency to loss of hair from the scalp. The first five symptoms are observable long before they would appear as simple results of inanition. In some of these cases, as in Mr. Skey's, the pain which follows the use of food is agonizing. I have a case in my mind of the last degree of severity, for her agony on the reception of no more than a cup of lithed broth was extreme. I mean that she would shrink away unnoticed and uncomplaining into any remote corner where she might moan to herself, unconscious of any observation. Now, to this lady, like all the sufferers of this class to which I have referred, the airs and humours of hysteria were wholly alien. Her daily life, even in the days of her weakness, was full of energy and of well-directed good works. Her habits were punctual and methodical, and her affections staunch and wholesome.*
She never vomited her food, nor have I found vomiting in any cases of anorexia nervosa. The chief varia-

---

* Laségue notices the briskness and energy of these patients, and gives a very fanciful explanation of them.

NEURALGIA OF THE STOMACH.

tion in the group of symptoms in these patients is in the presence or absence of pain. Pain, so acute in a few, is in most of them absent or insignificant. In the latter exists simply an invincible distaste for food. To yield to such a distaste may seem at first to many high-spirited girls rather a merit than otherwise; it may seem but a distaste for indulgence, a denial of the animal propensities. Thus creeps on, unknown or unconsidered, an aversion from food, and, indeed, an inability to take full meals, which seem, in some cases, almost unconquerable. If the fear of such a state becomes more prevalent among the public, it is probable that the known results of such mistaken self-denial will serve in future as a timely warning to prevent the full establishment of the distaste.

Hyperæsthesia of the stomach, then, may be found in anorexia nervosa, and to such cases is a grievous addition; in gastralgia of the periodic type, when it may concur with the attacks, or alternate with them, or again, it may itself constitute a form of gastralgia, a form in which pain occurs only during digestion. These cases are constantly confounded with dyspepsia; but the term can only be applied to them etymologically. For instance, No. 356, 1878, is a solicitor, aged thirty-one. He called upon me to complain of "grievous indigestion of long standing." Soon after meals pain would come on at the epigastrium, after a time generalizing itself about the chest or waist. It would last from one hour to two hours, and it never occurred on an empty stomach, not even if the due hour of a meal were overpassed. Now, this surely seemed due to some defect in the peptic process; and if it were not, how were we to distinguish its cause? Well, his tongue was quite clean, his secretions normal, and he had no offensive flatulence. Nor, again, did he much care what he ate; all things gave him pain alike, except that one thing, beef, which especially tries the neuralgic stomach. But fatigue or annoyance increased his liability to the torment. Sometimes, the pain had given place to a pain in the head, or sometimes it had "pinned him through the body." A year before this stomach-disorder set in, he "was demented with neuralgia in the head," and had suffered thus but little since. Before that, again, he had had a well-marked but brief attack of melancholia; his brother has hay-fever badly, and the family history is neurotic in other respects. Here then the gastric pain seemed to be of the hyperæsthetic form of gastralgia; nor was the diagnosis ultimately difficult unless we are urged to confound it with hysteria. No stomach-medicines could relieve such a patient, and this one was cured with Easton's syrup. In other cases, diagnosis might not be so easy as in his, and in some mixed cases it is very difficult to disentangle the pains which are caused by neurotic hyperæthesia from the pains which are caused by the presence of products of abnormal digestion. In Miss —— (No. 311, 1879) pain during digestion occurred as in the last case, but there were present also the characteristic nocturnal seizures. Miss —— (No. 31, 1880) suffered in like manner during digestion, but with great severity. It is strange to see that in her case the group of symptoms called anorexia nervosa never arose. She was pallid, emaciated, highly neurotic, and dreaded food. Still her pulse was rather quick, as well as feeble; the catamenia were regular, though scanty and watery; constipation was not well marked; and she manifested no loathing of food, but would gladly have eaten if she had dared. This lady had the abdominal aortic throb, and also, when the pain came on during digestion, her mouth would fill with clear water. It was a great pleasure to me to see this valuable life restored to health and activity after a long illness. A combination of morphia with arsenic and quinine, administered with food, did best service.

In Miss —— (No. 538, 1882) the pain only came on after distinctly improper food, such as tough beef, new bread, sad pastry or cake. Here is a transition towards mal-digestion. But the pain was so intense as to point rather to a neuralgia aroused by peripheral causes, and, moreover, pain was not unfrequently aroused by the same cause in the left orbit, or in the brachial plexus. The three pains might come separately, variously combined in pairs, or all together. This case reminds us of Watson's instance, in which a severe pain in the foot was brought on by eating an ice. I may indicate again the curious fact that such gastralgias may be associated with vomiting, or they may not be. In some cases a tendency to reflex motor spasm is there; in others it is not there. The former cases simulate ulcer, the latter do not, for I presume that vomiting is aroused in all grave cases of ulcer.

Miss ——, aged twenty-two (No. 416, 1883), came under me for intense pain after food, with vomiting and consequent relief. She was sent to me as a case of ulcus ventriculi, and I could not promise her at first that it was not. But her family history was neuralgic, and she used to be "a martyr to tic." Now, "the food makes her so sore and hurts her so," but she is relieved by vomiting. Whether the point be universally true or not, I cannot say; if it be, it is worthy of a pointed reference—namely, that we found warmth always gave relief in her case. I directed her therefore to drink hot water with her dinner, and after a meal to lie down under a rug, with a hot bottle to the epigastrium. This gave her much comfort, but it probably would not have relieved the pain of ulcer. On the other hand, in her, as in many like her, cold feet would precipitate an attack. This lady was completely and quickly cured by anti-neuralgic remedies.

Finally, we must remember, nevertheless, that, in organic disease of the stomach, the pain may be independent of food. Mrs. ——, aged fifty-two (No. 240, 1882), consulted me for recurrent pain, which was independent of food, and thus closely simulated simple gastralgia. There was no neurotic history, nor was the pain periodic. She vomited catarrhal matters; but some gastralgics vomit gummy phlegm with their attacks. The diagnosis was very difficult, until progressive emaciation and loss of complexion compelled us to infer the presence of carcinoma, an inference which was but too surely justified by the result. In another patient who was driven into this morphia habit by recurrent pain certainly diagnosed as gastralgia, adhesions indicative of old chronic inflammation were found in a limited district in the epigastric region not far from the surface. Unfortunately I have no notes of the post-mortem.

One symptom very characteristic of gastralgia next demands our attention, and that is flatulence. The flatulence of a typical gastralgic is a portentous thing. Like vomiting, like hyperæsthesia, it may concur with the pain, or come independently of it. It returns often in attacks as the pain does, striving vehemently, so that the concurrence of this wind and gastralgia make up the malady known as "the spasms." That such distensions and belchings as these are of other origin than the mal-digestion of food is, I think, very imperfectly realized by the profession at large. But in its more striking forms its independence of such causes is obvious. It recurs often with the same periodicity as gastralgia itself, and, like it, has a fondness for the small hours of the afternoon, or of the early morning, especially of the latter. The patient awakes, let us say, at 2 or 3 A.M., with a sense of distension which he knows but too well. His body is sometimes flat, but more often has swollen about the epigastric region, and become tight and hard. For a while, restless and distressful, he rolls upon his face, and rubs his belly until "the wind begins to move." This, as it does so, rumbles upwards rather than downwards, and delivers itself by no common eructation, but the working load rises in a series of eruptions, which roll from the throat in successive waves of flatuosity. It is a long storm rather than an explosion, and its tumult will keep the patient raised in bed, heaving in strife with it for an hour or two before he finds ease.

Miss ——, aged twenty-three (No. 233, 1883), migraineuse, is liable to violent attacks of flatulence. It awakes her in the night. She has no pain, but great distress; wind works upwards in enormous volumes. Her tongue is clean and tremulous, and the secretions normal. The gases ejected are innocent of all odour. The uterine functions are normal. It is not stated that this lady ever had gastralgia at all; certainly she had not at the time of my notes. A like state of things existed in a friend of my own, whose mental and bodily activities are incessant, and who had suffered greatly from gastralgia. He had consulted physicians high and low, home and foreign, his flatulence had invariably been put

NEURALGIA OF THE STOMACH. 61

down to "indigestion," and he was consequently dieted into still further emaciation, and still profounder gastric distress.

Miss ——, aged twenty-nine (No. 374, 1881), has gastralgia well marked, but complains as bitterly of flatulence. This always comes on upon an empty stomach, and is sure to be generated on the delay of any meal. It "rifts up voluminously."

Mr. H—— (No. 629, 1880) has had psoriasis for years. He is subject to attacks of "sinking in the stomach and bowels as if actually dying." He never had pain, but he has "rounds of intense and tremendous flatulence — always on the empty stomach."

Again, Mr. ——, aged forty-seven (No. 72, 1879), always "delicate," has gastralgia; at present daily. It recurs exactly at 4 P.M., and he hastens to relieve it by food. It is thus postponed till 7 P.M., when it is again put off by dinner. He is thus at rest till 4 A.M., when he is always awakened by "loud noisy and continuous belching . . . . with great rollings of wind and most painful distension." He presents no sign of disease anywhere, nor of disordered secretion. He never vomits: he has no discomfort from careful meals, but the reverse. He has had a continuous loss of blood from piles for a long time, and he looks anæmic. This fact he omitted to state until directly questioned, for he believed "that relief to be good for the system," and calculated to clear his "torpid liver." All these people are profoundly convinced that their main enemy is a vicious liver.

I will give but one case more; a solicitor, aged thirty (No. 73, 1878), who had been very hard worked. First came dull heavy pain in the left side. Then he became subject to "painful distensions of the body, not by food," and extreme flatulence; with this comes on great dyspnœa like spasmodic asthma, or again dizziness so severe that he fears he may fall. He passes profuse and pale urine at such times, and if a woman he would have been written down a hysteric: whereas he is a man of intellect and good sense, and holds successfully a high professional position. The only difficulty in his case was to secure for him a sufficient rest. It is to be noted that in all these cases the wind is quite inoffensive. Whether it be carbonic acid, common air, hydrogen, or what else I am unable to say, as I have never tested it. By its mode of occurrence, the suddenness and volume of its generation, and its innocence of taste or odour, it is clearly marked off from the foul wind generated by the fermentations of food. It is certainly produced in some way by the agency of the nervous system. Such are the painful symptoms which may be observed and discussed under the head of gastralgia; but a few yet remain.

There is a distension of the body, real or false, which sadly distresses some patients. Women having this distension must, perhaps, go upstairs to remove the corset for ease. The swelling is for the most part real and measurable; but I am not prepared to say what is the cause or mode of it. That it is a mere accumulation of wind, I hesitate to suppose. It is often unassociated with any consciousness of the presence of flatulence; and, moreover, it is not confined to the region of the stomach. I am assured by many patients, and their observing friends, that it may extend all over the breast, and even to the neck. It may be vascular, but there is no blush with it.

To the horrible death-like sinkings which belong to gastralgia, though not always actually coincident with the pain, I have already referred incidentally. No. 80, 1874, a male, aged twenty-eight, after a few months' worry began to have neuralgic pain in the left arm and up the neck. After this he became subject to nocturnal attacks of "want"—not a craving, nor exactly a faintness, but "an awful emptiness, with a dread at the heart." He was substantially quite healthy, and was soon cured on restorative and tonic treatment. I am now seeing a similar case in Leeds, with Mr. Robson, in a neurotic subject. In another case a gentleman, aged twenty-nine, who suffers severely from hay-asthma every summer, a craving sinking sensation, worse than actual pain, came on before meals. After meals, however moderate, he felt distended. He also discharged inodorous wind frequently from the bowels.

If he put a meal off beyond the usual hour the craving or sinking became worse and led to sensations of anxiety and exhaustion. His organs were all healthy, and he had been treated for "dyspepsia" for ten years. Vertigo is sometimes associated with the sinking as in No. 73, 1878, already quoted. Slighter degrees of this distress often accompany gastralgia or enteralgia, or, with an unaccountable languor, precede them. These subjective sensations are, of course, to be distinguished from the collapse, sometimes alarming in its degree, which is consequent upon the agony of the intenser attacks of abdominal neuralgia. The intensity, indeed, of most severe cases of gastro-enteralgia may be seen in the ashen cold face and blue nails. Finally, most gastralgics are subject, not when actually suffering only, but at all times, to borborygmi. Few people are ignorant of this phenomenon; but it is in neurotics especially that it reaches its most lively and garrulous form. An old and valued domestic, who has recently retired after twenty-five years of service, and who often presented all the phenomena of gastralgia in an active form, was so embarrassed by these audible internal questionings that she almost withdrew herself from waiting at table. In some persons borborygmi are rhythmical, and coincide with the inspiration: the noise is then rather of a churning character, and, I believe, is made in the stomach only. On examination, it will be found that in these persons the breathing, when full, is wholly abdominal. Even in women the thorax may be motionless. In addition to tonic measures gentle dumb-bell exercises are most potent to remove this disorder.

And now, gentlemen, have I succeeded in enforcing that which you already knew—namely, that all the symptoms I have discussed to-day form, not only a consistent, but a very uniform series ? Few maladies present a more definite series, or one more capable of rational explanation. Is it not one of the last to be grouped with those of irregular, capricious, and fanciful evolution ? Shall we then say, as the authors of our textbooks do to this moment, that gastralgia is for the most part a malady of hysterical women ? If we still say so, then it is indeed time for the lion to paint the man.

LECTURE III.

Mr. President and Gentlemen,—Enteralgia is, perhaps, the most racking torture contained in the curse of mankind; so bitter, so searching is the agony that more than one sufferer has said to me, calmly regarding his bygone pains, that surely the sting of death had been more welcome. That this malady is one of singular character and marked identity I had thought it my duty to show at greater length, until I turned but as yesterday to a description of it in Reynolds' "System of Medicine" by Dr. Wardell. In the famous and masterly handbooks of medicine which have been given to the profession of late years, the malady has received so scanty and indistinct recognition, it has been so often confounded with common gripes on the one hand, or contemptuously tossed over to the limbo of hysteria on the other, that Dr. Wardell's clear account of it is eminently valuable. Before reading this essay I had also noticed that the two chief points of onset are the umbilicus and the right iliac fossa, or, as I should put it, the umbilical region and the right flank. The central attacks are not strictly umbilical, but arise at a point usually about two inches to the side, generally the right side, of that mark; and the lateral attacks are generally a little higher than the iliac fossa, being nearer the level of the crest of the ilium itself. Another correction I offer is that females are not more prone to the disease than males. My own experience is of course inadequate to overthrow an assertion which Dr. Wardell says is beyond dispute, for, happily, the disease, if not actually rare, is far less common than gastralgia. Gastralgia, I have said, is commoner in women by two to one; but of fifteen marked cases of enteralgia, I find that eight occurred in men and seven in women. I should add that, out of a far larger number, I have selected the fifteen as severe cases uncomplicated with any intestinal derangement, or with any concurrent troubles elsewhere. The four most grievous instances were all in men of ages between twenty-two and forty-eight. Cases of mere tormina consequent upon irritation of the bowels or constipation, and cases of lead-poisoning, I have excluded; not because they have no element of visceral neurosis in them, as indeed they have, but because I desired to study the malady in its purer form. In this way we may hereafter carry our judgment more clearly to secondary or mixed cases. The greater relative frequency of enteralgia in men than of gastralgia may be explained in part by its being more directly allied to gout than is gastralgia, in part by its dependence on the weightier stress of the affairs of men. From my own cases I gather that a family history of gout, in its heavier and wider operations, is characteristic of enteralgics rather than of gastralgics. Enteralgia, like sciatica, may be a "pure neurosis," but is, like it, not rarely of gouty nature. I gather also that enteralgia is aroused rather by the strife of public life than by the teasing of homely worries. I will cite a few illustrative cases.

The Rev. —— (No. 54, 1883) is a minister of religion. He has the bright impetuous manner of a neurotic person, and has expended his energies lavishly in his work. He has suffered from no trace of melancholia nor of hypochondriasis; his knee-jerks and other reflexes are normal.* For twelve months he has been subject to attacks of pain in the right flank which are like the stabs and hacking of a knife. The agony is so frightful that he almost longs for his life to be taken. The collapse and subsequent prostration are very serious. Otherwise his health is good, and physical examination fails to discovery any disease of parts, or any morbid state of function. Dr. Daly of Hull had seen him and had pronounced the case to be one of neuralgia.

Another case of nearly equal severity was also in a clergyman, who made out for me a very graphic story of his malady. In him the seizure began, and

* It is better to state that I have discussed no case of visceral neurosis in which the knee-jerk was absent.

the abiding centre of the pain was fixed throughout all attacks, a little to the right of the navel. Intermittently, however, the pain would strike him "in a semicircular course, and settle in the kidney;" or, again, would dash from the centre "right through the liver and lodge under the right shoulderblade." In these seizures occurred, therefore, pains obviously spinal associating themselves with the enteralgia, as pains of the thoracic nerves may ally themselves with gastralgia. There was always incontinence of urine to a painful degree. "The liver and kidney tracks of pain rarely occurred simultaneously. The character of the pain at the centre was as though a spear-head was trying to force itself out from within." The attacks, in this case, have become rarer and slighter in the years 1879–83 than they were in 1876–78. The patient is a man of stirring life and of a lively and genial temperament. After endless medication for "duodenal dyspepsia" he improved on arsenic and on quinine with belladonna. One day, acting under some irresponsible advice, he took an overdose of "Mother Siegel's syrup," which prostrated him for a day or two; but after that event he had a long immunity from pain. There are adequate grounds for the belief of all his many advisers that he suffers from undeveloped gout. I may add that, in the earlier years of the pains, they were occasionally associated with ordinary gastralgia in the severe form.

In the case of Mrs. ———, a lady past the menopause, often sent over to me by Mr. Ball of York, severe enteralgia has accompanied, or of late succeeded, uterine neuralgia of a violent and intractable kind. All the parts of the pelvis and their functions have been normal. She has shown signs of indistinct gout for the greater part of her adult life. Her enteralgic pains were of a drawing or tightening character, but presented none of the ordinary symptoms of common colic. Sleeplessness and diarrhœa likewise afflicted her, and seemed to spring from the same diathetic cause.

Again, a medical friend writes to me of his own case, that pain in the right flank attacked him during sleep five years ago. It was not very severe, and was relieved by friction and a cordial. Since that time the attacks have been more frequent, more severe and more prolonged. Indeed of late he has never been free from some sense of pain about the same part, and, like a woman with a neuralgic uterus, he suffers especially if jolted in riding or in driving over a rough road. Of all remedial means hypodermic morphia alone relieves him. He has suffered from lithic acid in the urine in past years, but not of late; the regulation of diet and the use of alkalies having averted that disorder. There is no sign of any local disease.

In the case of a lady under the care of Miss Ker' of Leeds, enteralgic pains set in with a rough periodicity towards the early night. They are preceded by low spirits, irritability and a distressed expression of face. After times of ease the attacks have been recalled by fatigue or by exposure to cold. They may return nightly for a week or more, and then disappear for a while; perhaps for nine months little pain would be felt. Bed and hypodermic morphia alone give her relief. She is much thinner than she used to be. This lady was under my care for gastralgia some years ago. In a like case which I am now attending with Mr. Handcock of Leeds, the gastralgia of earlier life turned into enteralgia after the menopause. In her the attacks returned regularly at half-past seven in the evening. Enteralgia is rarely associated with vomiting, as is gastralgia; but the latter symptom is very prominent in a nurse whom I saw with Mr. Jessop a few weeks ago. It is doubtful, however, whether her case may not be of ovarian origin. Her catamenia are regular and normal, excepting some deficiency in quantity. Her bowels are quite regular, nor do I find constipation more common in enteralgia than we find it to be in many or most cases of atonic disorder.

In Case 320 (1875) the enteralgia was closely periodic, recurring about 3 A.M. during sleep. The patient was a gouty man. Gout also was a marked feature of the family history of a man, aged fifty-three, whose number is 371 (1875). No. 30 (1877), a man aged twenty-two, had suffered from the age of fifteen from alternations of gastralgia with definite umbilical enteralgia. Mrs. ——, aged fifty-six, (No. 269, 1877), became the subject of like enteralgia, after the cessation of migraine which passed away about the age of forty-five. The pain in her case was very severe. In another brother and sister of whom I have seen a great deal, enteralgia in a definite and violent form is the inheritance of the former, and a gastralgia of equal violence and definition attacks the latter. Both subjects are pure neurotics without gouty history. Miss —— (No. 508, 1881) was seen first by Dr. William Roberts, and by him her attacks were called neuralgic, as no doubt they were. The attacks were "like a corkstrew going into her flank [side not noted], and transfixing her." The pain often is so sudden in onset that she "stands like a statue, flushes, and then turns white, but has no loss of consciousness." Afterwards, "she writhes in fearful pain." There is no periodicity. Her hands and feet, as is usual in visceral neuralgia, are very cold, especially when the attacks are near. Her appetite, digestion and bowels are normal, as are all the functions of life. Her lips and gums are well coloured. She presents all the qualities of a neurotic, and her family history is markedly neurotic. Physical examination revealed no defects. This young lady two years later thanked me very sweetly for curing her. I find on turning to my books she had a course of bromides and arsenic with quinine, and pills of belladonna at intervals. After her recovery she had married, and continued well. Miss ——, aged forty-six (No. 618, 1881), had enteralgia arising near the navel. She is a neurotic person, otherwise healthy, and she lost her attacks by transference. She welcomed a series of violent attacks of right supra-orbital pain with muscæ volitantes, which took the place of the abdominal seizures. The attacks of Mrs. B—— (No. 639, 1881) were of the enteralgic type, but were followed by slackened bowels with a little mucus occasionally, though rarely. It is curious that in so many of my cases diarrhœa has been associated with gastralgia, and in so few with enteralgia. On the other hand, as I have said, I rarely find in the latter the constipation, the disordered alvine secretions, the flatulent colic which are described by other writers, I believe on grounds of probability rather than of direct observation. Flatulent colic is a wholly different affair; and, as a matter of speculation, I hesitate to place the seat of enteralgia in the bowel at all. In Mrs. S——, aged forty-three (No. 106, 1881), during attacks which correspond to the enteralgia of the other patients, the abdomen is said to become swollen and hard, the aorta throbs, and the heart palpitates; so that some general storm in the sympathetic accompanies the enteralgia, and the flatulence is probably due to that variation. As the attack ceases the abdomen falls, and it was quite soft and normal when I examined it. She was cured or relieved by bromides, belladonna and quinine. The case of Miss ——, sent to me by Mr. Knaggs of Huddersfield (No. 274, 1883), presents a good example of the gouty element. Her family history is gouty, and she had podagra at the age of seven. She was subject for many years to facial neuralgia; this left her for twelve months. Then came, and has often returned since, violent neuralgia in the subhepatic region. The attacks last three or four hours; and in her, by the way, some vomiting occurs, with little ejection. The attacks are brought on by fatigue, excitement or worry. She never had the faintest trace of jaundice. Rest and change relieve her. She suffers at times from gouty dyspepsia, but in no connection with the attacks. The pulse-tension is normal, the urine of specific gravity 1022; and all organs and functions are normal. Another very similar case was seen by me in the same year, gouty phenomena, migraine and facial neuralgia having preceded enteralgia. In a young man sent to me by Mr. H. Wright of Halifax violent enteralgia was associated with epilepsy, or rather was preceded by it. The enteralgia showed a very remarkable periodicity in this case. A curious series appeared after great mental strain and anxiety in a patient whom I saw for enteralgic seizures with Mr. Holmes

of Leeds. First came cervico-occipital neuralgia
with intense hyperæsthesia of the scalp, loss of sleep
and appetite and great debility; then a strange
sensibility to cold upon the skin, so that the lightest
draught of air distressed him, even the turning of
the leaves of his ledger; then periodic coryzas, as
intense as if dilute ammonia had been poured over
the mucous surface, an abundant fluxion being
followed by general prostration; this developed into
frequent asthmatic seizures, called forth by cold
drinks, or by the least change of temperature in his
room, but cut short at once by coffee. These symptoms had been relieved by sea-water baths before
my visit. I must not linger over these cases, however, many of which do but repeat the symptoms
already described. I shall, however, refer to four of
them hereafter, which are associated with symptoms
on the skin.

Not included in the list now discussed, but natu-
rally following it, comes a series from which I need
not quote largely, if at all; but I may indicate, as
not uncommon visitors, certain patients who complain
of a pain in the region of the hepatic flexure of the
colon, and I will go farther, and say I believe it to be
seated in that place. It is not severe nor periodic; it
is rather wearisome and abiding. It is found in the
melancholic rather than in the brisk neurotic; but,
like enteralgia, seems clearly associated with gout.
Although not connected with obvious disorder of the

bowels, yet a course of blue pill and Carlsbad water generally relieves it—a course which would drive an enteralgic subject into a frenzy. The pain of lead colic, though probably of immediate nervous causation, is nevertheless not difficult to diagnose from the preceding forms. It has no limited *lieux d'élection*. It is more diffused than enteralgia, and more extended than the pain of melancholia just described. It is periodic, doubtless; but it is rarely an excruciating, rarely a slashing pain; it is rather of a grinding or fretful character. The parallel constipation and other circumstances will, I believe, always serve to distinguish it in those few persons who, suffering from lead-poisoning, have no blue line.

On turning to consider the neuralgias of other viscera, we are met by the difficulty, as yet insuperable, of proving definitely in what nerves the pain may lie. We know that all essays to localise abdominal pains in the splanchnic nerves and their paths to the sensorium, in the sensory fibres around the large arteries, or in the deeper connections of the spinal nerves, are mere gymnastics, barren exercises which may stimulate the fancy, but which as yet can neither be proved nor disproved. We can only find our way through those obscure gates and alleys of the body by patient clinical watchings, by grouping and contrasting our cases, and by noting the curious warps of disease. Are there such affections, for instance, as hepatalgia, nephralgia and so forth? Careful clinical observers, such as Anstie and Spender, say there are such affections; and with them I am disposed to agree. I have careful notes of three cases in which neuralgia seemed to me to be seated in the liver, so far as I could judge by the situation of the pain.

Mr. W. A——, aged thirty-two (No. 656, 1881), whose health is habitually good, whose habits are temperate, and who presents no obvious disorder of function, has called upon me at intervals for three years. Four months before his first visit he was taken with a pain which he precisely refers to the seat and extent of the liver. This pain has often recurred, observing no period of recurrence, save that it always attacks him in bed at night. It is a "miserable pain;" he rises and paces the floor for hours. He maps out the liver, of whose seat he was previously ignorant, with curious exactness. He has had no jaundice, nor does he suffer from constipation. The pain does not stab nor radiate, as spinal pains would do. On bromides and arsenic he recovered, and was well for twelve months, when worry and overwork recalled the attacks. The family history points to rheumatism. I have about five cases in my books similar to this, all agreeing with Dr. Spender's description of typical cases of hepatic neuralgia, in which the patient suffers from severe pain of the neuralgic type, deep in the region of the liver, with no inflammatory symptoms, and with intermissions of perfect ease. In some cases a characteristic shoulder-pain is present also. Hydrochlorate of ammonia, he says, will cure it.* Anstie's words are to the same effect, if a little less confident; and he also marks as characteristic a peculiar mental depression which accompanies the pain. Now, I wish especially to point out that the case which Anstie quotes is of a girl of eighteen, who suffered from repeated attacks of the kind, and that after one of them a jaundiced tint appeared, and rapidly disappeared. Subsequently, the pains appeared again without the colour.†

Let me parallel this with the following. Miss—— (No. 14, 1882), aged twenty-two, consulted me for severe pain in the hepatic region passing towards the stomach. I was disposed to regard the attacks as gastro-duodenal neuralgia. There was decided evidence of neurotic tendencies in the family history, and in her own a liability to nervous diarrhœa. She presented no signs of disordered function. Her complexion was as clear and pretty as her features deserved, and she showed no marks of anæmia. I fear I did her but little good, the attacks recurred, and she sought other aid. But here comes the point of the case—namely, that on some later date she passed a gall-stone with the usual symptoms, and shortly afterwards she passed a second. Beyond this point her history is unknown to me. These facts I

* "Relief of Pain," p. 125.   † "Neuralgia," p. 62.

had from her own medical man. Little as one could dream of gallstones in a young and clear-skinned girl—in one, too, whose alvine disorders tended rather to looseness than to constipation, and whose habits were alert and active; yet there they were. Again, I saw some months ago, with Mr. Field of Dudley Hill, a woman of later middle age, whose sufferings were hepatalgic. She had never shown any trace of jaundice, the pains were of a neuralgic type, and she was of a neurotic constitution. Yet, on deep and close manual exploration, both Mr. Field and myself became convinced that there were gall-stones palpably in the gallbladder. Another case I saw with an old pupil and friend, in which like unaccountable hepatalgic pains recurred, and greatly distressed the patient, also a middle-aged woman. She died some weeks later of peritonitis due to the perforation of a gall-stone, which was extracted after death. Dr. Anstie evidently took the jaundice in his case to be neurotic; but, put with my own, does not the group of them prove, or strongly suggest, that hepatalgia is, in many cases at least, a sub-hepatic pain arising in the gall-bladder and the ducts thereof, and due either to the irritation of gall-stones or to some vicious quality of the bile? In this way I have explained other cases I have often come across, and which at one time puzzled me—cases in which pains of hepatic and epigastric origin, but without jaundice, recur frequently and violently at times between attacks which are obviously due to the travelling of gall-stones. At present, then, I regard hepatalgia as a pain aroused by the coincidence of an impressionable or neurotic habit with the presence of gall-stones at rest in the bladder. That pain may not arise in these tubes independently of such or similar irritations I will not, of course, assert.

The same reflections occur to me when I pass on to consider nephralgia. Anstie suspects that many cases of so-called nephralgia have been cases of pain lying more outwardly. In such a sceptical spirit I minutely searched into the following case. A lady, aged forty-one (No. 528, 1881), was sent to me by Dr. Charles Smith of Halifax. Two months previously, after great harass, she woke at 5 A.M. with intense pain in the loins, chiefly on the right side. It was dreadful pain; her aspect became death-like, her finger-nails black and her lips ashen. Thus she suffered for three mortal hours. The pain also ran round the front of the abdomen on the same side and down the groin into the thick of the thigh. Her micturition was frequent for some hours, and it then fell away to scantiness and rarity. The same series of pains recurred every morning at the same time, though happily with less severity, and slight degrees of it were felt occasionally at other times. Both Dr. Smith and myself sought in vain for any gross cause for such attacks; the only discovery made was of some grey sandy matter of an indefinite kind in the urine. Dr. Smith says, " I never could satisfy myself that there was a calculus in her kidney. When she gets below par she still has returns of the renal pains with the same flow and ebb of the urine. I suppose," he adds, "a visceral is like a facial neuralgia, which when due mostly to a condition of nerve, is called a pure neuralgia, when mostly due to a condition of tooth is called toothache, but in a large mass of cases is caused by a condition of nerve not bad enough to show itself if not irritated by a condition of tooth. In this category comes Mrs. ——'s case, a nephralgia due not entirely to nerve, but coming partly from the kidney having to perform abnormal work. I wonder whether a neuralgia of the kidney could produce swelling of the testicle, not the epididymis. I have seen a case which I could only interpret thus." I have quoted Dr. Smith's remarks in full, as they apply so well to the mixed cases of gastric and other neuralgias I have tried to interpret on like principles. That in Mrs. ——'s case the conditions of nerve predominated seems probable, because she was a highly nervous woman and had suffered from other forms of neuralgia, because the attacks showed periodicity, and because no other considerable cause was discoverable. The small quantity of fine gritty matter seen occasionally in the urine seemed, however, in some of the attacks to be the *causa efficiens*.

Another case is No. 384, 1883. Mr. ——, aged twenty-seven, has suffered at times for three or four years from pain in the region of the right kidney, and at times very severely. The pain is then a continuous ache, with exacerbations; it runs, when more acute, down the tract of the ureter to, or nearly to, the bladder. At times its severity causes cramps of the limbs and collapse. There was never a trace of blood, and the differential diagnosis from calculus was carefully considered, and decided against calculus. The urine was wholly negative, not even unduly acid. He improved upon quinine and belladonna.

No. 394, 1882, was a commercial traveller, aged thirty-nine. He suffered precisely as No. 384; and the two cases received much attention from me, as they came within a few days of each other. This patient had also neuralgia of the left orbit and cervico-occipital neuralgia with Valleix' points. There was no syphilis, by the way, in any of these cases.

Before dismissing the subject of the kidney, I may remark that, in chronic renal disease, in granular kidney especially, gastralgia may appear with its craving appetite, sensations of want and so forth, as in the uncomplicated form. I think this has been spoken of by previous writers, and so also has the concurrence of gastralgia with aortic regurgitation. This coincidence I had failed to verify until recently, when three well-marked instances came under my observation.

Dr. Ralfe has kindly forwarded to me the following account of a form of renal neurosis often associated with valvular disease of the heart.

"In 1878, the late Dr. Murchison gave me the particulars of a peculiar form of neurosal attacks, which he designated as 'renal storms,' and which had frequently occurred in a patient suffering from aortic regurgitation. The attack commenced with excruciating pain over the region of the right kidney, exactly like renal colic, but there was no sickness nor retraction of testicle; the urine passed during the attack and immediately afterwards was perfectly normal, nor was there any jaundice to suggest that the pain was due to biliary calculus. After lasting some hours, the attack passed off as suddenly as it came on.

"In 1880, I saw a man, aged forty-seven, who applied as an out-patient at the London Hospital, solely on account of a severe paroxysmal attack of pain, which, commencing at the angle of the epigastric region where it joins the right hypochondrium, passed deeply downwards into the right lumbar region. No disease of the liver or kidney could be detected, and the urine was normal. On examining the chest, the heart was found diseased (aortic regurgitation).

"The only reference bearing on this form of neurosis or 'renal storm' is in Dr. Habershon's 'Lectures (Lettsomian) on Diseases of the Liver,' page 13, in speaking of the neuralgic pain sometimes met with in organic disease of the heart, and which is referred to as being deeply situated behind the first part of the duodenum: 'It is severe, almost like the pain of gall-stones, but it is without jaundice or other symptoms of calculus; it is not connected with the stomach, for it is not affected by food, but paroxysmal and recurring sometimes with great regularity.'"

I will not pursue these reasonings so far as to consider in like manner the pains of other viscera, such for instance as of the bladder and the rest, for the same principles apply to them all. In many cases a neurotic tendency may be aroused by a local irritation which, in robust persons, would be almost, or quite, unfelt; or the neuralgia may be as "pure" as face-ache may be. Neuralgia of the rectum, however, I must not wholly pass by, for it is a definite malady, not very uncommon, and seems almost always to be "pure neuralgia." It is rarely very violent, being more of a thrusting, aching pain. It is mentioned by Anstie and other writers; I have met with it occasionally, and Dr. Myrtle made it the subject of a paper lately read to the Yorkshire branch of the British Medical Association. It is generally easy to distinguish from the pain of fissure and other local defects.

DIATHETIC AFFINITIES.

An allusion to neurotic diarrhœa, a somewhat different phase of neurosis, is again but a passing duty. Neurotic diarrhœa is a troublesome affection which, I think, is tolerably well recognized. Many persons owe the frail health of a lifetime to its recurrence, for it is sadly incurable, recurring and again recurring, uncontrolled in obstinate cases by anything less than opiates. Milder cases may be regulated by bismuth and the bromides, followed by pernitrate of iron or sulphate of iron pills. In this disease (which must be distinguished from catarrh and from diarrhœa mucosa), the stools, though slackened, are rarely watery, and rarely contain much mucus. They may or may not be attended with pain; they are reproduced by nervous causes; they are generally worse in the earlier day, especially before breakfast, and in women they cling rather to the catamenial periods. Arsenic is not often useful in these cases, and, indeed, can rarely be borne. The ailment occurs in both sexes, falling, perhaps, equally upon both. The diarrhœa is generally associated with other neuroses, such as migraine, cardio-vascular instabilities and the rest. It is found at all ages, for it often begins in early adult life, and is seldom wholly removed. The pathology of the day will attribute it, no doubt, to vaso-motor disturbance, and compare it with polyuria, lacrymation and night-sweats. It may be the homologue of pyrosis.

I would now pass on to consider the genetic

affinities of the class of patients which we have discussed in these lectures. This question is more easily answered by searching among the gastralgics, as they form by far the larger number.

One fact strikes the observer very early in his work; and that is the frequent association of neuroses of the vagus with certain kinds of eczema, lichen and psoriasis, with eczema and lichen chiefly. Of the alliance of gastralgia with asthma I have already spoken; and the frequency of eczema in asthmatics is generally recognized. Now eczema and its allies are nearly as common in gastralgics as in asthmatics, so that there seems to be some sympathy between the vagus nerve and cutaneous nerves. The skin affections and the gastralgia, if concurrent in the family, may not concur in the individual, or, if they do, they may not be contemporary in him, nor, if contemporary, parallel in severity. No. 245, 1883, a man, aged thirty-eight, came to me four years before for boils; at a later date, he came to me for gastralgia of the ordinary type, without functional stomach disorder. For this I gave him arsenic, and cured him. Last year he came to me to tell me for the first time that he had suffered for six years from eczema on and off, and that the medicine he took for the gastralgia had removed also the eczema. On this occasion he called to ask for a renewal of the prescription, not for the gastralgia but for the eczema. Again, No. 226,

1882, male, aged forty-seven, had excellent health till ten years before, which he attributed to the periodic returns of " eczema and psoriasis." Of late years he had seen but little of the eruptions, but his health was broken down by "chronic dyspepsia," for which he had been treated in vain. I found no signs of functional disorder of the stomach; the symptoms were those of ordinary gastralgia, and were cured quickly by arsenic. I find a history of skin affections in about eighteen per cent. of my cases as they stand in my notes, and I suspect this estimate, if we were to take a survey of families, would be found under the mark. As in No. 245, 1883, skin affections, if not actually present, escape record even in the individual life, unless especially inquired into. It has been fancifully assumed, therefore, that gastralgia is an eczema of the stomach, an assumption with as little of probability as of proof. I have not, however, heard that any writer has called asthma an eczema of the lungs. Are we then led by way of cutaneous eruptions to a recognition of gout as the spring of these, of gastralgia and of asthma? Not directly, I think, though I lean to the suspicion of some remoter affinity herein. I find that experience justifies the distinction between the eczema, lichen and psoriasis of "dartrous," and those of "gouty" habits. The distinction is partly to be seen in the histories of the patients and of their relations, and partly, though not invariably, in the characters of the eruptions themselves. In the dartrous, no doubt, these eruptions tend rather to be more acute and more symmetrical ; to moisten, to spread and to itch greatly ; while, in the gouty, they tend rather to imitation, to dryness, to lack of symmetry and to a less active tint of hyperæmia.

The dartrous diathesis of Bazin and others has, then, gastralgia among its evolutions, as the gouty may have ; but the term dartrous may well be put in a subordinate place, and the term neurotic diathesis used to indicate the family. Healthy persons with dartrous eruptions, if followed home, will often turn out to belong to neurotic families. Two brothers of my acquaintance illustrate this. The one is a very healthy man, with the exception of a continual liability to eczema ; the other is an asthmatic. There are other strong neurotic characters in the previous generation. But lately, again, an odd thing happened to me. A young man, seemingly very healthy, came to me for severe gastralgia of a pure type. I pressed him as to skin eruptions, and he denied their presence in himself and in his family. Within the next six months there came to my rooms first, his sister, a healthy-looking girl with recent and vicious-looking eczema around both ears ; then the mother with symmetrical moist spreading eczema of the legs ; and, finally, the baby of another daughter with eczema of the head. These successive cases were almost put down to my ominous inquiries, and, in an earlier century, I might have been burnt. There was no gout in the family, nor any suspicion of its latent presence. Another diathesis which I find often recorded in my cases is rheumatism— rheumatism of the kind which culminates in rheumatic fever.

Miss ——, aged seventeen (No. 350, 1876), had eczema capitis between the ages of seven and eight. She was then, and later, a somnambulist. After this she had " horrible sinkings" in the forenoon for some time, and was treated for "dyspepsia." These vanished in favour of intense neuralgia of the head. In 1875 this again gave way to attacks of gastralgia, so severe at times as to end in a faint. She was treated again by several medical men of position for " dyspepsia," and her diet corrected until it became spare and monotonous. She grew so much worse that she was taken to Sir William Jenner, who reversed all the rules of her dietetics, fed her liberally, and gave her strychnia. She recovered in a few weeks. She called upon me for a new trouble—namely, for intense neuralgia of the left orbit, often leading even to ecchymosis of the conjunctiva. Now in this young lady's family rheumatism and rheumatic fever were well marked on the paternal side; her father was a delicate man and a martyr to rheumatism; he had had rheumatic fever more than once.

Another such case occurred in Miss ——, aged

twenty-five (No. 480, 1880). She came of a neurotic family, and one from which phthisis had not been absent. Her consultation was in respect of gastralgia of an intense form. She had been migrainous most of her life, and she had suffered from rheumatic fever at eighteen years of age.

Now, do these cases, which I can strengthen by many others, lead us to make any direct tie between the rheumatico-gouty diathesis, the eczema and the visceral neuroses? No, I think no direct tie. To turn aside for a moment, I have to set before you another affinity, which must be considered with those. Perhaps the disease most largely found in neurotic families beyond their neuroses is phthisis. In case after case I find phthisis in parents or kin. This applies to all the neuroses, not to the gastro-abdominal only. I may refer again to the last-named case, No. 480, and, farther, to Mrs.——, aged thirty-five (No. 128, 1878), who came to me for a uterine ailment. She had had four children. I found some catarrh, both uterine and vaginal, the uterus soft and the surrounding parts relaxed. She had the usual dragging aches of the pelvis, pain on jolting, tenderness to the exploring finger and so on. She is liable to severe neuralgia in either supra-orbital notch. Although an abstinent woman, she is sick on rising in the morning, and all her symptoms come more about her with the approach of the catamenia. These return every twenty days or thereabouts. She has gastralgia very often, with voluminous flatulence and pyrosis. In her family there is a strong phthisical history. She made a good amendment on sedative and tonic treatment with change of air, and she has continued to enjoy better health. Her uterine troubles vanished as she regained strength, a course of douches only being prescribed.

Mr. H——, aged forty-five (No. 403, 1883), has suffered acutely from pseudo-angina, but consulted me for enteralgia arising near the umbilicus. His family history is badly phthisical.

Mrs. —— (No. 414A, 1883), aged thirty-two, has had severe facial neuralgia, but now consults me for enteralgia and urticaria. She has had attacks of eczema. Her uterine functions present no special derangement. Her family history is very phthisical.

But I need not multiply cases which present like features merely to record that phthisis belongs to their series. This I may invite you to observe for yourselves. But, for many years, I have noted the frequent coincidence of phthisis and rheumatic fever in family trees, and I think this coincidence has been observed by others. To my mind, then, the natural group is neurosis, phthisis and acute rheumatism. But what of the gout? Rheumatism and gout are certainly akin, the two maladies appearing in the same families, and even in the same persons at different ages. Yet I think gout goes by a different route to its direct work. A pure neurosis goes hand in hand with phthisis and acute rheumatism; but pains which, like migraine, enteralgia, sciatica and so forth, may, in such association, be of the purest type, may, in other instances, be caused by the poison of gout, and thus should be put into a separate class with the eczema and psoriasis of gout, which, again, are not "dartrous." In the one class the neuralgias depend upon some congenital defect of nutrition, affecting the nervous apparatus itself; the thing is essential: in the other class the nervous apparatus, originally healthy enough and even vigorous, is angered by a circulating poison. Yet is there not some alliance beyond this? Is there not some remoter pathological kinship, unexplained by the humoral theory? Yes, I believe there is. I believe that, farther back, there is some community of nature between the gouty and the neurotic habits; and to this we may ascend when we track out those phenomena which suggest that gout is itself originally a neurosis. Although there is, then, no cross alliance between neuroses, phthisis and gout, yet one may trace the history of gout upwards to a point of union with a divergent stock, which, two, three, or four generations back, had branched off into the tuberculo-neurotic. In such cases, and I have notes of several of them, anti-gouty remedies are of no use whatever. Finally, there is another form of neurosis of gouty origin, for the relief of which anti-gouty medication is useless; and that is seen in skin affections or neuralgias which had been directly aroused by the active presence of gout, but which were left stranded after the gouty humours had ebbed. Flotsam and jetsam of this kind are common in practice, and in such cases we do ill to run upon specific medication. A hemiplegia may remain, forgotten as it were, when the cerebral centre has healed; and will persist, perhaps, unless driven off by faradism. A syphilitic ulcer may remain after the syphilitic state has been wholly counteracted by the usual remedies, and decline to heal until some more general treatment is taken up. In like manner, an eczema or a neuralgia may be set up by gout, and this, as it ceases to be active, leaves behind it some altered or lowered state of tissue which no longer acknowledges the first disturber of the peace, but lies in a simpler dejection of its own. We must try the specific measures; but if these fail we must not therefore assume that our surmises as to the original cause of the ailment were false. These are the remnants which are sometimes suddenly routed by a blister or an acupuncture. Urticaria is not an uncommon addition to other neuroses, as is well known, and I forbear to record many of my cases in which it was found together with visceral neuroses. One I have cited above, under No. 414A, 1883. Zona is, perhaps, never found with a visceral neurosis, unless there has been some cerebro-spinal nerve disturbance therewith. Mrs. W——, aged sixty-six (No. 546, 1381), came to me for general and nervous debility. She told me she had suffered repeatedly from angina pectoris (whether the true angina or not I cannot say), but with the last attack there broke out a violent crop of shingles, after which the angina returned no more. I saw a remarkable series of this kind in a patient of Dr. Steele of Morley. She was a woman somewhat beyond middle life, whose whole temperament and history afforded a good picture of the tougher kind of neurotic. When I saw her the neuralgia of the viscera had extended to the cerebro-spinal system, and so involved the nerves of the left arm that a crop of zonal vesicles occupied almost the whole length of the ulnar aspect of it from shoulder to wrist. There was no gouty history in the case.

And now, gentlemen, my limits warn me that if I am to comment upon the treatment of visceral neuroses I must hasten to that end. Although the basis of all therapeutics must be a clear diagnosis, yet, on the other hand, the most elaborate diagnoses are but laborious idleness if not made the means of cure. We cry out with the child in La Fontaine,

"Eh! mon ami, tire moi de danger
Tu feras, après, ta harangue!"

The beginning of all successful treatment must be to convince the patient of the true nature of his malady. Now, your neurotic is one who has no reserve. This want is probably due to a congenital instability of nerve, showing itself as waste so ceaseless that the reserve once dissipated is never re-accumulated. This reserve may have been spent in beneficent activities; or it may have been dissipated in fidgets, fretfulness or shrewishness; in sleeplessness, in anxiousness or in pain, according to the quality of the person. We are disposed to forget that the silent work of nutrition uses more force perhaps than many people expend in their neuro-muscular life; hence the early failure of the digestive resources in neurotics, hence the fall of the balance of nutrition below the needs even of a controlled expenditure. We know that good nutrition is the main source of steady work, good temper and self-control; we know, too, that to trade daily only upon the supplies of the day is to court collapse; we must have more brain, more spinal marrow, more liver, more kidney than we want for the day. We must have stored-up force, partly for greater occasions, partly to secure the equable running of our machinery. A neurotic person is an engine with a light fly-wheel and a small furnace, whose work, therefore, is fitful and unsteady. In the early life of neurotic subjects, especially, we must husband the reserve, we must control expenditure, and cherish nutrition. To heap up again a wasted reserve is always a long and laborious task, and, as years go on, becomes harder and harder. Many people, even when under middle age, never wholly replace their reserve if wasted, let us say, by acute disease; so that this factor of steadfastness and safety is wanting in the work of all their later lives. In an exhausted neurotic the secret of treatment, therefore, is—by food, fresh air, exercise and happiness—to lift up your patient from the invalid-couch, not for the day or for the month only, but to teach him so to manage expenditure and so to promote nutrition as to replace his capital. He must establish in himself the habit of a cool temper and rhythmical work, and root out the habit of wasting volumes of good worry upon forecasts of events which never happen, and on visions of unsubstantial things. Unhappily, however, nine neuralgics out of ten are possessed by the belief that they are dyspeptics, and that the cure of their malady is to be found in some farther defalcation of their diet, in some new arrangement of it, or in the use of some new combination of stomachics. Once convince them that the stomach is the seat of neuralgia, and that any pain caused by food is as accidental as is the increase of pain in tic douloureux during the act of mastication; once let them realize that, so far from waving their dishes aside, like the physicians of Sancho Panza, you would rather prompt them to indulge as liberally as their impaired forces will permit; once persuade them to throw all their alteratives, their pepsines and their mineral waters to the dogs, and your battle is half won. I cannot read the lives of men like De Quincey or Carlyle without suspecting that a timely or untimely course of Fowler's solution would probably have deprived us of the letters of Mrs. Carlyle and of the "Confessions of an Opium-Eater." Assuredly a gastralgic can no more eat a good dinner than he can walk a league, but by careful advances, and the repetition of small light, highly nutritious and social* meals, he will eat his way upwards. The first dread of food over he will begin to digest *con amore*, and he may no doubt be helped to this end by the aid of pepsine, though my patients seem to do nearly as well without it. In severe cases, the warmth of bed for a few days, or even for two or three weeks, is of great value; economy of work and economy of heat being thus secured. Many a case of neuralgia which had resisted all medicines has been cured by a course of bed alone. The very common association of cold extremities with gastralgia, and, indeed, the oft-noted influence of chilled feet in setting up an attack, is an indication of the need for equable warmth. Dwindled meals fail to supply both heat and nutrition; and, in extreme cases, the Weir-Mitchell system, by stimulating nutrition, may re-

* A very distinguished man of letters, of nervous temperament, once told me that in good company he could eat an equally good dinner, but dining alone he had to limit himself to the simplest fare.

open the fountains of warmth and vigour. The visceral neuroses again, like the rest, are aggravated by cares and sorrows, and by depressing conditions of life. Release from toils or worries, and a change of air and scene, take a leading part in the cure. Gastralgics as a rule do better inland than at the seaside. At the sea they are apt to become irritable or sleepless; but these ill effects are lost on withdrawing a mile or two inland. On the other hand, except in the case of young anæmic people of fair physique, the high mountains are not favourable to neurotics; they fare ill at St. Moritz and Davos. Milder upland airs like Malvern or the Sussex Downs (dry sunny slopes a little remote from the seashore) suit them best. For how many patients can we write the prescription—to take two months' holiday, to withdraw from all toil and care, and to live in good company on refined and delicate food? And yet how potent are such means when all else may fail! Mr. Teale and myself tried in vain to cure a clever, impulsive and overworked Leeds clothier of gastralgia, mixed with some consequent dyspepsia, until we sent him on a sea-voyage. His own words on his return were: "After a few days bile could not live in my stomach, and my tongue was as clean as if I had manufactured it myself." The only things necessarily to be forbidden are tea, coffee, tobacco, and the stronger meats, such as beef. Over many persons thus susceptible tea has an evil power; it retards the pulse, and creates the horrible empty exhausted feeling, which seems as hard to bear as very pain. Alcohol I do not encourage in neurotics; that there is a little occasional help in it, I admit; but, on the whole, alcohol, drawing as it does upon the reserve fund which we wish to protect, is better away from persons who may learn to take it rather as a dram than as a small addition to meals; this error, in them, is a radical one. A like danger may follow the use of morphia, but severe cases cannot be treated without it. The repeated attacks so exhaust the patient, that it is only by economizing his forces with warmth, rest and morphia that he can retain any for the absorption of his food. Morphia may be given fractionally in ordinary mixtures, or periodically in larger doses; but in either case the remedy should be kept under the control of the doctor; in many cases it is even well to keep the patient in ignorance of the agent. For this reason I often order Dover's powder in pills, in order that the compound may not be recognized in the prescription as opium. Of other drugs, arsenic in gastralgia takes by far the chief place; indeed, it is hard to say how gastralgia was cured before the time of its introduction by Dr. Leared. Yet, even now, its power is not sufficiently well known, for, on turning to Dr. Ross's work on the "Nervous System," which I suppose to be the best in our language, I find no record of arsenic as a remedy for gastralgia; and Dr. Spender's rules for the use of the drug are too timid. Yet, after all, with soft nutritious food, warmth, rest and lenitive or narcotic doses of opium, many cases of gastralgia still resist treatment. Curiously enough, a repetition small blisters to the epigastrium may then be of service; and, of other drugs, quinine, boldly pushed on, with belladonna, forms a valuable combination; and so, again, do quinine and bromide of ammonium dissolved in hydrobromic acid. The infusion of the Prunus Virginiana makes an excellent vehicle for such mixtures. Not uncommonly gastralgia is a product of malaria. Of this nature I have two cases before me, and in one of them very large doses of quinine cured a most intractable gastralgia, which had resisted all other measures. Luckily, I knew my friend had travelled in the East, and had ague there. The silver salts, again, are of undoubted use; with nitrate of silver I cured one case which had defied all my previous measures. Of manganese I have no experience. Of iron I speak last; it has only been of much use to me in a few cases, for I do not, in fact, observe that anæmia, apart from the general lowering of all nutrition, has been so frequent a feature in my cases of gastralgia as many authors definitely assert of their own. Where any distinct anæmia exists, iron, of course, is indicated, and often works a cure. Phosphorus is not so useful in gastralgia as its kinship to arsenic would lead us to expect; but the pharmaceutical compounds of the hypophosphites now sold do, by virtue of some one or more of their constituents, seem at times to answer well. As the stomach gains vigour cod-liver oil should be added to the dietary, it will help on nutrition and forward recovery. In a word, arsenic and quinine are the only specifics; and the rest of the treatment may be summed up in rest, sedation, nutrition and tonics. Some gastralgics find that alkalies give them a temporary relief from pain, even in cases of neurotic and periodic type. It is not generally so, however, and the practice is not a sound one. When we leave the vagus nerve, when we leave asthma, angina pectoris and gastralgia, we find that the specific powers of arsenic are no longer so trustworthy. In enteralgia it may have some value, but far less than in gastralgia; in enteralgia quinine and belladonna seem best to forward restoration, though arsenic is, even here, by no means to be despised. In all visceral neuroses a most careful search must be made for any kind of peripheral or humoral irritation, and such irritation soothed and its causes averted. Of the infinite pains, moral and dietetic, which are needed for vomiting cases, I need not speak, for the management of them is sufficiently well known. The only unfamiliar drug which I can recommend for these cases is the walnut-spirit* sold by Messrs. Corbyn and others; this medicine, which was indicated some time ago in the *Practitioner*, I have certainly found very useful in cases of neurotic vomiting. It must also be remembered that gastralgic vomiting is spasmodic asthma of the stomach, and that a few whiffs of chloroform, or a little subcutaneous morphia, may cut the one short as well as the other. Finally, my remarks would be incomplete did they not contain some reference, however inadequate, to the percussion-cure of Dr. Mortimer Granville. Surprising results seem to have been obtained by this method in certain cases of deep as well as of superficial neuralgias, but the time has scarcely come yet for a confident judgment upon its merits.

Mr. President and Gentlemen, I must not detain you longer. Happily free as I am from neuroses in general, yet a sinking at the heart has possessed me many a time as I wrote and then read the lectures I have ventured to present to you. How slender is the offering none knows better than I. One while I have comforted myself with the thought that the late Dr. Symonds did not think the subject of headache

---

* A tincture of walnuts made by steeping the nuts in brandy is an old domestic remedy for stomach complaints. It is to Dr. Edward Mackey, of Brighton, to whom I am indebted for a more definite knowledge of this agent; his paper was published in the *Practitioner* for 1878.

too trivial for your thoughts ; but again I reflected how much the more must be the talent of the speaker the less striking his subject, and how great the interval between him and me. That I have said nothing new to you I am painfully aware ; that my words may not have fallen below the authority of this chair is my single hope.

THE END.

www.ingramcontent.com/pod-product-compliance
Lightning Source LLC
Chambersburg PA
CBHW031405160426
43196CB00007B/902